Pervasive Systems and Ubiquitous Computing

PERVASIVE SYSTEMS AND UBIQUITOUS COMPUTING

A. Genco and **S. Sorce**
University of Palermo, Italy

A. Genco and S. Sorce
University of Palermo, Italy

Published by

WIT Press
Ashurst Lodge, Ashurst, Southampton, SO40 7AA, UK
Tel: 44 (0) 238 029 3223; Fax: 44 (0) 238 029 2853
E-Mail: witpress@witpress.com
http://www.witpress.com

For USA, Canada and Mexico

WIT Press
25 Bridge Street, Billerica, MA 01821, USA
Tel: 978 667 5841; Fax: 978 667 7582
E-Mail: infousa@witpress.com
http://www.witpress.com

British Library Cataloguing-in-Publication Data
A Catalogue record for this book is available from the British Library

ISBN: 978-1-84564-482-6

Library of Congress Catalog Card Number: 2010920130

The texts of the papers in this volume were set individually by the authors or under their supervision.

No responsibility is assumed by the Publisher, the Editors and Authors for any injury and/or damage to persons or property as a matter of products liability, negligence or otherwise, or from any use or operation of any methods, products, instructions or ideas contained in the material herein. The Publisher does not necessarily endorse the ideas held, or views expressed by the Editors or Authors of the material contained in its publications.

Printed in Great Britain by MPG Book Goup, Bodmin and King's Lynn.

Contents

Preface

The ancient Greek agorà was the place where people met other people to communicate or discuss philosophical issues as well as human daily troubles and joys. Nowadays, we are still attracted by the same kind of place even if in a new virtual modality which is now made possible by internet technology. The new current agorà has different names, for instance Myspace, Facebook and other virtual squares where we go to when we want to encounter real or virtual friends, or want to shop in a virtual market place.

The internet agorà has broken generational walls so that older people, as well as the young, want to spend part of their own time with a computer and internet applications.

The only troubling side of that is in considering a computer something like a medium totem where we need to go, or a window open into the main virtual square. Although many people feels it very comfortable to stay at home and interact with others worldwide from one's own beloved armchair, the pleasure of going outside and meeting real persons and shops should not be in contrast with internet services. Ubiquitous Computing and Pervasive Systems are novel compromises which are capable of putting together internet services and real open environments. All that we do by means of a pc, we can now do also living and moving among real people and real things, with a little help from wireless technology.

Ubiquitous Computing and Pervasive systems are no more futuristic visions; they are something easy to be implemented. Mobile devices and programming languages are there, available to be used to this end.

The question why pervasive applications have not fully replaced pc internet applications yet is very likely to be singled out in commercial issues. Nevertheless, the pervasive solution does not seem to have actual alternatives at the moment, and it seems more likely to have the strength of an obligatory direction.

Many engineering faculties introduce pervasive systems in regular courses as well as other faculty, as for instance in the field of motor sciences or commerce, where the actual advantages of a pervasive technology are as evident as attractive.

This book has been written mainly having in mind its use as a text book for regular courses in engineering-technological faculties where a wide discussion and technical elements are requested.

A. Genco, 2010

Chapter 1

Introduction

Pervasive systems implement a middleware paradigm to make Mark Weiser's vision real. In 1988 he entitled 'Ubiquitous Computing', a project to be started at the Computer Science Laboratory at Xerox PARC. His idea was that 'The most profound technologies are those that disappear. They weave themselves into the fabric of everyday life until they are indistinguishable from it'.

According to Weiser's vision, a computer we interact with should not be monolithic or bounded in a place. Computer should be used to provide us with an artificial extension of the reality we live in and we interact with. A so-called augmented reality can then take place to be perceived in the same way as we are used to manage our reality in everyday life, at any time and in any place, inside or outside home or work.

Ubiquitous computing (UC in the following) is also called as pervasive computing and is based on the integration between computer processing and common-use objects by means of very small micro systems whose presence we cannot detect or we are not interested to detect.

Unfortunately, when Weiser expressed his vision, micro system technology and wireless communication were not so effective and affordable. His idea was considered just a scientific hypothesis and it was ignored in practice for a long time.

Only from 2000 we can detect some confident researcher reflection on the possibility of using currently available technology and actual implementation of UC. In particular, wireless communication's lowering costs pushed factories to invest in pervasive systems for home automation, as well as for Internet services, by the use of mobile devices such as notebooks, personal digital assistant (PDA) and cellular phones.

The transition towards Weiser's vision is currently underway, and we can predict that in a few years many surrounding objects will be provided with micro systems and embedded autonomous processing capabilities.

According to new suitable modalities, we will find us interacting with an augmented reality. Customized visits to heritage sites, real-time pervasive medical assistance, car navigation, in-field automatic training, mobile commerce and so on are just some examples of facilities, already partially available today, that will give us several advantages. These will be, at the same time, exciting as well as potentially worrying. There is a risk that we may get a new digital dependency which will further complicate our life. Not less worrying the idea of letting pervasive systems hold some personal data of ours, it will be needed to enable pervasive systems to provide us with services exactly customized for us. Therefore, we will be often called to choose between keeping our privacy and enjoying pervasive services.

UC is mainly implemented by means of pervasive systems, those systems which work spread over the environment and use the environment itself as an interaction medium between people and computer networks.

Pervasive systems' basic concepts are those of augmented reality, that is, the reality enriched by virtual contents, and of disappearing hardware, that is, of hiding hardware from our perception. Computer is no more visible because it is hidden by the environment veil and because in the near future we shall not have the need to bring a computer with us to enjoy computer services.

Information processing systems become an integrating part of reality, with the role of providing reality with artificial intelligence behaviours. Environment's physical and conceptual reality is then enriched with entities which are designed to emulate intelligent behaviours.

From the human perception point of view, the environment in augmented reality can be interpreted as being aware that human–environment interaction takes place not only according to the rules of a natural reality but also to the ones requested by some kind of distributed artificial intelligence. We shall be called to familiarize with new types of feedbacks which will not be so predictable as the ones of physical lows and nature physiology. A feedback will come from some artificial intelligence process which will be arranged according to a programmer's logic.

This aspect can excite some perplexity, but, however, the opportunity for programming environment feedbacks allows us to arrange interaction systems and mediate human expectations with programmable virtual entities.

Augmented reality is then the environment where hybrid entities live. They are partly real and partly virtual, some kind of wrappers with the capability of combining an artefact or a human being with some individual knowledge dealing with it.

Computer services become contextual or context-aware. They perform according to modalities and contents that depend on context elements such as who, where, when and why. When we use a context-aware service we perceive something different from any another person who might use the same service. According to what is written in some personal profile of ours, we feel just the elements of reality that fit our interests or wishes. There will be as many digital realities as we are. It will depend on us whether to take a look on other's realities.

Once provided the environment with suitable digital equipments, pervasive systems implementation can be carried out basing on hybrid entity working models. Human–computer interaction then evolves and becomes human–environment interaction up to augmented human–human interaction.

This book deals with those models and technologies that put together the bases of pervasive systems. On a conceptual plane, we deal with *disappearing hardware* and *augmented reality*. For technologies, we spend some words on *wearable computer* and *wireless communication*. And for principles of operation, we investigate on *positioning*, *security*, *human–environment interaction* and *service discovery*. Throughout this book we will take into consideration some applications of the ones we can find in literature, for instance, those for heritage sites and for positioning.

Chapter 2

Augmented Reality and Virtual World

1 From virtual reality to augmented reality

1.1 Virtual reality and virtual world

Virtual world (VW) is a world with artificial elements all generated by computer according to virtual reality (VR) techniques. VW is made of virtual entities that are programmed to stand as functionally equivalent to real ones.

VR turns out very effective to simulate environments uneasily accessible in real world. It is low cost, risk free and safe, and it is capable of implementing huge environments, as in astronomy, as well as very small ones, as in the case of cellular systems. VW can simulate dangerous action sceneries for flight, fire, contaminated areas and so on. VR allows us to face and solve problems which are often without other solution and gives us the chance of getting operation abilities or observing simulated behaviours.

New VWs could include abstract entities, such as logical elements, reasoning, grammatical and syntactical forms, thus enabling us to face the ambit of word and thought. It could include psycho-motor faculties and define new interaction types up to new alternative communication forms.

VR can manage economic-financial values and turns out very useful when it is difficult to guess how a share may behave in selected conditions. It will be much easier to study a share trend in a virtual environment that is specifically arranged to observe its interaction with other shares or to formulate a new understandable model.

Dealing with art, VR can create virtual relationships between space and music to manipulate sounds in space rather than in time, or it can manage pictures and let us navigate into a hypothetical 3D projection.

1.2 Augmented reality

Augmented reality (AR) is something more than VR; virtual entities become hybrid entities and can include and manage real objects. Their main task is to mediate interaction between real and virtual resources.

As Azuma [1] says in his Survey of Augmented Reality 'Augmented Reality (AR) is a variation of Virtual Environments (VE) or Virtual Reality as it is more commonly called. VE technologies completely immerse a user inside a synthetic environment. While immersed, the user cannot see the real world around him. In contrast, AR allows the user to see the real world, with virtual objects superimposed upon or composited with the real world. Therefore, AR supplements reality, rather than completely replacing it'.

AR allows all that can be generated by a computer to be superimposed on physical reality, from simple description data up to complex virtual structures in VW.

Reality is then enriched with contextual contents as well as with new models and contents that can develop from invention.

2 AR technologies

AR technologies are partly the same for VW because in both cases we have the need for advanced visualizers and immersive environments. As far as visualizers are concerned we can use

- HHD: handheld display,
- HMD: head-mounted display,
- SID: spatially immersive display.

In VW, these devices allow a user to see only computer-generated virtual images. In AR, the same devices allow a user to see virtual objects superimposed on real objects.

2.1 HMD (head-mounted display)

There are three main types of advanced HMD systems: optical HMD (OHMD), video HMD (VHMD) and retinal scanning display (RSD). All displays can be monocular or binocular. A monocular display shows images to one eye (Figure 1), while binocular displays show the same image to both eyes. In this case, images are a bit different to achieve a stereoscopic view and let human eyes give a volumetric representation of the observed objects (Figure 2).

The main advantage of HMD is that user hands are free and can operate comfortably. The disadvantage is that users cannot see anything else than

Figure 1: Monocular HMD.

Figure 2: Binocular HMD.

what is projected by a computer in the display and therefore they cannot see anything of the surrounding environment.

2.2 Optical HMD

An OHMD uses an optical mixer which is made of partly silvered mirrors, and it has the capability of reflecting artificial images as well as letting real images to cross the lens and let a user to look at the world around him.

Optical mixers do not let light completely reach user's eyes and therefore real-world view turns out to be veiled and virtual objects seem to float on the real ones rather than hiding them.

Furthermore, due to image contrast being affected by lightening intensity, digital graphics are unlikely to be easily distinguished in a bright environment or in front of a light background.

Another problem comes from the different focus planes of virtual and real images, so that it is quite difficult to have both images properly focused [1].

Generally speaking, an optical display cannot perform like a human eye either for space or for colour.

2.3 Video HMD

OHMD problems are partly solved by VHMD. VHMD devices merge artificial graphics with the images coming from video cameras mounted on them. A merged view is then projected on the display which is totally opaque.

Virtual and real images are perfectly synchronized by delaying real images by the time taken by a scene generator. The main problem of such devices is to correctly position the cameras so as to give a correct stereoscopic view to human eyes. Cameras are obviously mounted in places other than eye's position. Therefore, a parallax error takes place and it can cause users feel unwell when they stop using a VHMD and resume looking at the real environment around them directly with their own eyes.

A z-buffer algorithm is used to manage the *z*-axis in a 3D system. This turns out very useful when overlapping real and virtual images and lets artificial images partly or totally overshadow real images behind them. This way, a realistic view is then achieved where real and artificial images are correctly rendered.

As in the case of OHMD, VHMD also cannot give the same resolution of human view, and it becomes very difficult to make shadows exactly fall from virtual objects on a real environment.

The AR goal is to perfectly mix two worlds, real and virtual. Image rendering by means of HMD is, therefore, very important to let us feel both worlds as one (Figure 3). Unfortunately, even a few pixels deviation between the two representations are detected by human eye, and it can heavily affect the vision of an augmented world. There are some techniques available indeed to face the problem that, however, can affect image fluidity.

Figure 3: An example of augmented environment with perfect alignment (on the right), where the digital augmentation of cups, floppy disks, pencils and rubbers can be noted.

2.4 RSD (retinal scanning display)

RSD scans the light rays coming from an object into a raster structure, pixel by pixel, directly to the optic nerve through the cells in the retina of the person who is looking at the object. This gives high-quality digital images, wide and bright, without additional displays.

RSD is made of one light emitter, one modulator, two scanners (horizontal and vertical) and optical components [2]. The light modulator regulates the photon stimulation intensity of retinal receptors. The scanners are synchronized with the light modulator by means of some electronic video components.

The optical components magnify images and make the scans converge to achieve an optical focus [3].

RSD can be very effective in AR. Its direct projection gives the highest image definition and a wide visual field. A RSD device is very light and cheap. Display brightness is directly regulated by the scanned ray and, therefore, it can be used even in bright environments as, for instance, a sunny place.

2.5 HHD (handheld display)

HHDs are unlikely to be compliant with AR philosophy and need drawing our attention more to the reality around than to the devices we use.

HHDs can be considered as advanced personal digital assistants (PDAs). They are equipped with LCD display and embedded video camera to get real video superimposed by digital images.

In spite of their non-compliance with AR philosophy, HHDs can help AR diffusion, because their technology is already available in PDAs and smart phones which are likely to be equipped with AR software in the near future.

2.6 SAR (spatially augmented reality)

A real environment can be augmented by integrating virtual objects in it. For instance, digital projectors can create images on walls and digital displays can be installed anywhere.

The main advantage of SAR, in comparison with HMD and HHD, is the capability of freeing users from carrying unusual hardware with them, thus leaving people to immerse themselves in AR and interact with it.

Several SAR systems have been tested so far with 2D and 3D floating images projected on planes or irregular real objects. Some tests were carried out with one projector, others with more than one. The simplest one is with one projector without any concern with user position. SAR suffers from alignment problems when 3D images are projected from different projectors; however, these problems can be solved by using calibrated video cameras.

2.7 SID (spatially immersive display)

SID is a typically large display which enriches an environment with digital images. An SID system is usually arranged with multiple front or rear projection display.

One of the most known SID is CAVE (which is the recursive acronym of cave automatic virtual environment), a system which uses rear projectors to arrange 3D immersive environments [4]. A stereoscopic view is achieved by alternating images for left and right eyes which are adapted according to viewer's head position. A subsequent simplification is Immersadesk [5], which uses table size displays, giving a good degree of full interactive immersion [6].

Further advancements allowed designers to implement tile displays which were arranged by placing several displays one beside the other. ActiveMural is an example of these systems which are mainly aimed to reproduce huge environments under normal light conditions. μMural is a mobile solution which implements boundary merging for better image quality. Finally, InfinityWall is based on only one high-definition big screen, which avoids the need for big control panels, thus making user interactions with the system simple and natural.

2.8 Augmented tools

AR can be also pursued by enriching usual tools with some processing capabilities in addition to their original feature and usage mode.

An example of such AR kind is MediaCups, an ordinary coffee cup which is invisibly augmented with sensors, processing unit and communication devices [7]. In the same category, we can find smart labels for door, clocks and everything that fantasy can suggest.

Augmented objects in the same environment can communicate, interact and share environmental digital information between them. Daily use object computer augmentation will not evolve to look like computers; they will preserve their original features in addition to what embedded digital processing can do [8]. Of course, augmented tools can suitably work only in augmented environments which they can interact with.

3 AR and VW projects

It is often hard to say where VW ends and AR starts. They share common boundaries where the two ideas are undistinguishable. We could assert AR is artificial and then not true, while VW always stands as real elements even if it is a reality abstraction.

As far as AR techniques are concerned, there are many ranging from databases to identification systems.

Generally speaking, on the basis of information stored in some remote server, a system tries to recognize the environment and identify the real objects in it by reading digital data such as pre-recorded videos or simple tags attached to an object in the environment. Reading is made possible by some wireless communication technology like, for instance, infrared (IR) or radio frequency (RF) (Figure 4).

Other systems, especially those for external, try to identify the environment and objects inside, by localizing a user and using a server-provided map of the information to be displayed in each place. An example of these systems is mobile-augmented reality system (MARS) that was designed to provide a user with a virtual guide [9]. The user equipment in MARS original version is rather bulky. All the needed devices such as notebook, batteries, pointing devices and a positioning system with 1 cm accuracy, also capable of detecting and measuring head motion, are put in a 13-kg kit-bag to carry.

3.1 MediaCups

The MediaCups project [7] applies AR technologies to a coffee cup along with sensors to detect the position and rotation (Figure 5). Furthermore, it uses an accumulator that is wireless recharged by means of the saucer and it communicates through IR with a network infrastructure.

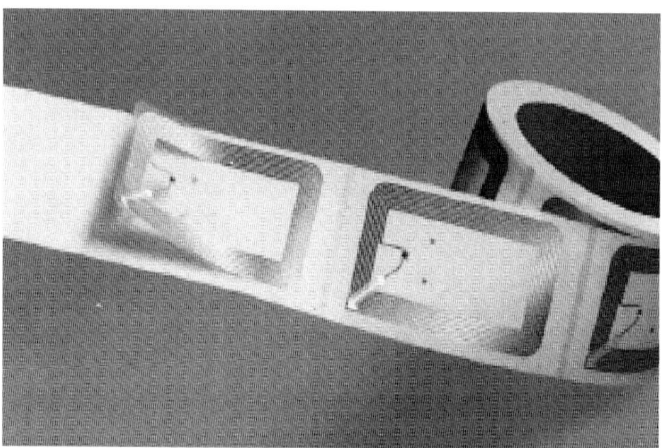

Figure 4: Adhesive RFID tag.

Figure 5: A cup in the MediaCups project [7].

3.2 ActiveSpaces

ActiveSpaces [10] uses a set of wide displays, for instance, the InfinityWall, thus creating an immersive space, intuitive and natural, where any interaction with digital systems can be simplified and expanded by AR.

The initial project was developed by the Future Lab group at the Argonne National Laboratory of Chicago University. Next, it became a research topic for many other corporations and institutes worldwide.

The ActiveSpaces project operates on a working environment by combining existing infrastructures with new advanced computer technologies. The goal is to arrange augmented working environments that give the feeling of entering a shared environment also connected to other working environments. This can allow research groups to investigate on huge data spaces by an interactive and visual way.

3.3 Access Grid

Access Grid [11] is complementary to ActiveSpaces, where it enhances its communication section. Access Grid is meant to be a resource set aimed to facilitate interaction among remote working groups.

According to the grid computing concept, with multiple computing resources in a network aimed to give a shared and integrated environment, the goal of Access Grid is to allow users to arrange distributed cooperative environments, called nodes, so as to help interaction among multiple working groups.

Access Grid pursues its goal by starting a high-performance video-conference in a location suitably equipped for group work as for both installation and logistics.

The location effectiveness of course depends on the availability of audio-video equipments, such as wide displays, and software for sharing ideas, applications and discussions.

The ideal cooperation environment will be a space designed to satisfy some main requirements. It must gratify who is inside, and it must entail a feeling of co-presence with other groups who are using similar spaces.

In addition, as far as remote interactions are concerned, the active space will reflect human communication modes as, for instance, the way of engaging somebody in conversation, frontally, privately or whispering, as well as all the mental conditioning occurring when we are in a group in a same place.

Acknowledgements

This chapter was written with the contribution of the following students who attended the lessons of 'Grids and Pervasive Systems' at the faculty of Engineering in the University of Palermo, Italy: Scrima, Di Trapani, Lo Cascio, Failla, Meli, Fichera and Sangiorgi.

References

[1] Azuma, R.T., A survey of augmented reality. *Presence: Teleoperators and Virtual Environments*, **6(4)**, pp. 355–385, 1997.

[2] Tidwell, M., *A Virtual Retinal Display for Augmenting Ambient Visual Environments*, Human Interface Technology Laboratory, Washington Technology Center, University of Washington, Seattle, WA, 1995.

[3] Johnston, R.S. & Willey, S.R., Development of a commercial retinal scanning display. *Proc. of Helmet- and Head-Mounted Displays and Symbology Design Requirements II Conf.*, 18 April, Orlando, FL, pp. 2–13, 1995, DOI: 10.1117/12.209726.

[4] Cruz-Neira, C., Sandin, D.J., DeFanti, T.A., Kenyon, R.V. & Hart, J.C., The CAVE: audio visual experience automatic virtual environment.

Communications of the ACM, **35(6)**, pp. 64–72, 1992, DOI:10.1145/129888.129892.

[5] DeFanti, T.A., Dawe, G. & Sandin, D.J., Immersadesk, http://www.evl.uic.edu/core.php?mod=4&type=1&indi=163, retrieved on June 2009.

[6] Disz, T., Papka, M.E. & Stevens, R., UbiWorld: an environment integrating virtual reality, supercomputing, and design. *6th Heterogeneous Computing Workshop (HCW '97)*, p. 46, 1997, DOI: 10.1109/HCW.1997.581409.

[7] Beigl, M., Gellersen, H.W. & Schmidt, A., MediaCups: experience with design and use of computer-augmented everyday artefacts. *Computer Networks*, **35(4)**, pp. 401–409(9), 2001.

[8] State, A., Livingston, M.A., Garrett, W.F., Hirota, G., Whitton, M.C., Pisan, E.D. & Fuchs, H., Technologies for augmented reality systems: realizing ultrasound-guided needle biopsies. *Proc. 23rd Ann. Conf. on Computer Graphics and Interactive Techniques*, pp.439–446, 1996.

[9] Höllerer, T., Feiner, S., Terauchi, T., Rashid, G. & Hallaway, D., Exploring MARS: developing indoor and outdoor user interfaces to a mobile augmented reality system. *Computers and Graphics*, **23(6)**, pp. 779–785, 1999.

[10] Childers, L., Disz, T., Hereld, M., Hudson, R., Judson, I., Olson, R., Papka, M.E., Paris, J. & Stevens, R., ActiveSpaces on the Grid: the construction of advanced visualization and interaction environments. *Parallelldatorcentrum Kungl Tekniska Högskolan Seventh Annual Conference (Simulation and Visualization on the Grid)*, Lecture Notes in Computational Science and Engineering, eds. B. Engquist, L. Johnsson, M. Hammill & F. Short, Springer-Verlag: Stockholm, Sweden, Vol. 13, pp. 64–80, 1999.

[11] Argonne National Laboratory, Access Grid. http://www.vislab.uq.edu.au/research/accessgrid/, retrieved on June 2009.

Chapter 3

Human–computer interaction

1 Introduction

1.1 Definition

Human–computer interaction (HCI) is a discipline devoted to the design, evaluation and implementation of interactive computing systems to be used by the humans. The main focus is on the evaluation of interactive systems and on the analysis of the phenomena arising from their use.

The heart of the matter is the concept of 'interface' between man and machine. The true meaning of the term 'interaction' is varying in this respect, because 'man' and 'machine' can have different interpretations in different application contexts.

HCI studies both sides of the interaction: systems and humans. As a consequence, HCI is a multidisciplinary field that involves computer science (design of applications and their interfaces), psychology (application of theories on cognitive processes and empirical analysis of the user's behaviour), sociology and anthropology (interaction between technology, work and enterprises) and industrial design (interactive products).

The main concerns of HCI are the execution of cooperative tasks by humans and machines, the communication between man and machines and human capabilities in using machines (interface comprehension, interface usability, ergonomics and so on).

The study of communication between humans and computers relies on several disciplines such as computer graphics, operating systems and programming languages on the computer side, while social sciences and cognitive psychology are involved on the human side.

1.2 HCI and ubiquitous computing

In the past decade, the main goal of the research in the field of ubiquitous computing (UC) has been the support to humans in their everyday life without upsets.

In the Weiser's vision, a great variety of heterogeneous devices with regard to their shape or use will be at the disposal of the user [1]. Such devices will be aimed either to be 'personal' or 'embedded' in the environment.

Some common UC devices are personal digital assistants (PDAs), tablet personal computer (PC), laptops and electronic dashboards. The research activity is oriented to devise more and more sophisticated objects and to their integration through unwired communication channels.

1.2.1 Classic HCI

Till 1960s, when the computer began to be a commercial product, the user was not an autonomous subject in the interaction with the machine. Computer users were go-betweens from the actual stakeholders interested in data analysis to the machine used to elaborate data. The computer was also passive because it did not posses a 'face', there wasn't a screen yet.

Cathode ray tube (CRT) video terminals were introduced in 1971; they replaced printers as the main output device of the computer allowing to present information very quickly.

Nevertheless, the interface layout was poor on the screen also. Interfaces lacked in ergonomics; they were difficult to learn and use and only a few specialized people could use them. Moreover, different applications had different interfaces, thus making more and more difficult the learning phase in using a software package.

PCs appeared in 1981. It is the bridge between 'centralized' and 'distributed' use of computing systems. When using a PC, the user has a private interaction with her own system that is completely under her control. Moreover, floppy disk and hard disk technologies allow storing a virtually unlimited amount of information.

The PC entered the everyday life as a common device for unskilled users also. In turn, a growing need arose for more simple and efficient computer interfaces than in the previous decades.

At first, the use of the menus involved the possibility to assign a user's choice to each row. Commands could be issued very simply by checking the positions in a menu through a cursor. Menus introduced a double information coding: both 'spatial' and 'verbal'.

The introduction of the mouse increased the importance of spatial information coding. The user can now execute actions directly on the screen objects, thus obtaining a sensory feedback (visual or auditory). Each logic action corresponds to a physical one, producing a tangible effect.

After menus and mouse, interfaces were enriched with 'icons'. Icons are small graphic objects that are used to indicate the user's choices. Each icon is associated to an action of the user, and it is arranged to symbolize in a very expressive way the effects of performing such an action. Icons are used as physical objects to be manipulated through the mouse. With the introduction of icons, true graphical user interfaces (GUI) arise that are based on the window-menus-icons-pointers paradigm (WIMP). GUIs are designed as metaphors of everyday life objects to stress their learnability. The most famous metaphor is the 'desktop metaphor' that was invented at the Xerox Labs. Starting from the mid of 1980s, Apple spread it out all over the world.

Along with technological improvements, the first theoretical studies on HCI are conducted during the 1980s. At the beginning of 1990s a sudden change took place with regard to the way of designing a GUI. The interface is now a user-centred system. Moreover, the first analyses on the target users of particular computer systems are carried on.

Interface design is now a true discipline that is faced in a scientific way, with the aim to support humans in the task of facing complex problems in the surrounding reality.

1.2.2 Modern HCI

Nowadays, the meaning of HCI is related to man–machine interaction that takes place in social and organization contexts where different systems are intended to satisfy different human needs.

In this field, humans are analysed according to

- their psychology,
- their abilities and
- their physiological limits.

In a nutshell, HCI involves the communication between humans and computers and their abilities when using complex systems such as interface learnability and performance measures in task execution.

Starting from 1960s all these topics have been deeply analysed by the various scientific disciplines involved in HCI. At first, human information processing during the interaction process was studied with the aim to build a 'model of humans' to be useful during the design phase.

The first usability tests were developed at the beginning of 1980s along with the growth in the use of PCs. Technological developments during the 1990s have supported a strong use of the HCI concepts in modern workstations via an increase in the computing power, the communication bandwidth and the graphic devices (touch screens, virtual or augmented reality and so on).

Nowadays, the research in the field of HCI is oriented towards the use of computers inside a workgroup, the so-called Computer Supported Co-operative Work (CSCW), media integration and 'multimodal interfaces', and the effects of new technologies in working and domestic environments.

2 Implicit and explicit HCI

2.1 Implicit and explicit HCI: a comparison

When considering current computer technology, interaction is explicit: commands are issued to the computer using a particular abstraction level (command prompt, GUI, gestures, and written or spoken natural language). In implicit interaction, the user performs an action that is not intended to be a computer command, but it is 'interpreted' by the machine as an input.

Implicit interaction relies on the computer's ability to understand human behaviour in a specified context. Let us consider as an example a computerized trash can that is able to read bar codes of the items the user throws away and suggest the shopping list accordingly.

The user performs a simple action that would be the same regardless of the garbage can. Moreover, the user does not interact with computer that is embedded in a particular can, so the whole process describes an implicit interaction. The example points out that implicit interaction is based on two main concepts:

- perception and
- interpretation.

In general, implicit interaction is used in addition to the explicit one. Three main concepts can be devised that facilitate implicit interaction:

- the user's ability to perceive the usage, the environment and the circumstances;
- some mechanisms to understand sensors percepts and
- some applications that are able to use such information [2].

2.2 Implicit HCI

In human communication, most of the information is exchanged implicitly. Often, implicit contextual information like posture, gestures and voice intonation ensure the robustness in man-to-man communication. Moreover, body and spoken language are redundant, i.e. nodding one's head and saying 'yes'. Such implicit knowledge is used to make the information more clear.

Applications can be improved by implicit HCI. The application I/O and its execution environment have to be analysed to this aim; in turn, the context of use has to be evaluated along with the application feedbacks.

2.3 What is 'context'?

The word 'context' has a lot of meanings in computer science, according to the particular research field like natural language processing, image understanding, computer architectures and so on [2].

A complete definition of 'context' in HCI can arise by focusing on the following five questions:

- *Who*. Current computer systems focus the interaction on a particular user regardless of the other ones in the same environment. As human beings, we adapt our activities and remember past events depending on the presence of other people.
- *What*. Interaction assumes knowledge about the user's action. Perceiving and understanding human activities are very difficult tasks. Nevertheless, a context-aware system has to face them to produce useful information.
- *Where*. Spatial information is a relevant part of context, in particular, if it is joined with temporal one.
- *When*. Most of the context-aware applications do not use time. Changes that take place over time are very interesting to understand human activity. As an example, if a user spends very little time on a particular screenshot, maybe the user is not interested in what is displayed. Moreover, actions that are far from a particular behavioural model can be relevant for the system. As an example, an interactive domestic environment should be aware of an elderly person who does not take their medicines.
- *Why*. Understanding why people perform actions is more and more difficult than understanding what an action means. A good starting point is the use of context information like body temperature, heartbeat and galvanic skin reflex to obtain information about the emotional status of the user [3].

2.3.1 Context representations

Defining the notion of 'context' implies a model to represent it. Good general representations of context do not exist, so the application designer has to develop ad hoc schemes with limited capabilities to store and manage the information described earlier. 'The evolution of more sophisticated representations will enable a wider range of capabilities and a true separation of sensing context from the programmable reaction to that context' [3].

2.4 Explicit HCI

Explicit interaction takes place by means of visual interfaces where information is coded through 'icons'. In this respect, the designer has the goal to build an interface enabled with visual consistency that allows the user to make sense of it. Widgets, icons, windows, menus and all the components of a GUI represent the moves of a dialogue between the user and the system. They have to be designed following some 'usability principles' (refer to [4] for a detailed explanation of this topic). In a GUI, the 'point-and-click' metaphor replaces the command line interface where the user has to learn a suitable language to issue commands by digitizing them on a keyboard.

Modern usable interfaces have to obey three main principles:

- learnability,
- flexibility and
- robustness.

The general principles can be articulated in several facets that define in detail the requirements of a usable interface.

Generalizability is a crucial usability principle that can be regarded as a particular facet of learnability. Generalizability refers to the interface's capability to allow simple interaction in a new application context based on the knowledge of similar situations in other applications. Generalizability is strictly related to consistency, that is, the ability of the interface to exhibit a similar behaviour to other applications when the user has to face similar tasks. Consistency can be referred to whatever feature is in the interface, and it is often measured in terms of the I/O behaviour.

Another important usability principle when designing an interface is its *recoverability*, that is, a particular facet of robustness. Recoverability is the ability of the users to recover from the errors when they detect them. Recoverability allows the users to exert control over the task they are carrying on through interface because they are able to come back from an erroneous interaction path and they are always prompted about possible errors related to the next move.

Finally, the designer has to take care of those interface's components that carry a particular meaning; such elements have to be designed in order to draw the attention towards them. One can use a particular arrangement of shape, colour and position to achieve this goal.

The concept of usability has been standardized by ISO, even if usability principles have not been coded as design rules, and they are the result of several best practices in the design process [5]. A designer has to follow them to obtain a good interface that allows the user to have a good dialogue with the system.

3 Adaptive and intelligent HCI technologies and methodologies

This paragraph is devoted to artificial intelligence (AI) technologies and methodologies used in adaptive intelligent interfaces for UC devices [6].

The general adaptation process is described in Figure 1.

The main modules in the Figure 1 are

- perceptive processing,
- behavioural processing,
- cognitive comprehension and
- adaptive interface.

In this framework the user interacts with an interface that changes with respect to the mental state. The *processing* modules receive as input some pieces of information regarding the user's sensory status.

The *perceptual* module processes images of the user's face and eye (eye position, pupil size and so on). Moreover, it analyses the body and its postures.

The *behavioural* module analyses the actions performed on the interface: pressed keys and the force exerted in pressing them, mouse movements, clicks number and position, and so on.

Both modules provide data that are merged in a unique sub-symbolic representation to allow a learning algorithm to recognize some emotional states like confusion, stress, mental fatigue and so on.

Symbol grounding to perceptual data has been discussed for a long time in several disciplines like cognitive sciences, AI, computer vision and robotics. Percepts can be represented as a set of signals that are continuous

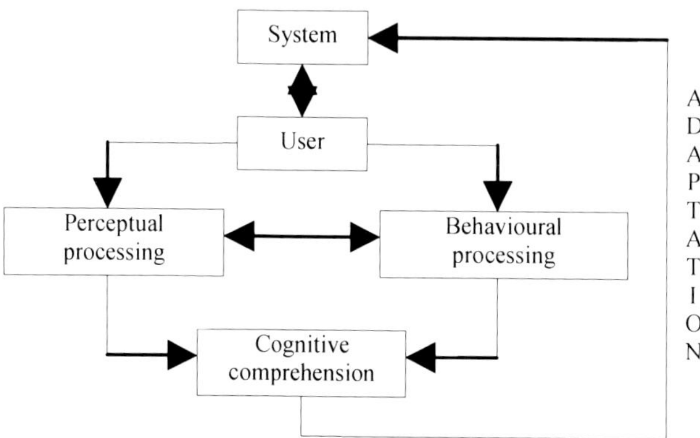

Figure 1: Adaptation process in AI that is used for intelligent interfaces.

both in space and in time (heat, pressure, sound, eye perception and so on). Often, behavioural strategies are elaborated with regard to a change in environmental conditions that in turn can be detected as discontinuities in such signals.

Refining the techniques for perceptual and behavioural processing is the key for a good design of adaptive interfaces along with the use of effective cognitive modules. In this way it is possible to obtain personalized GUIs by detecting automatically the user's features.

Feature extraction is a model-driven process. The model starts with an initial configuration that is adapted by detecting the displacement of the body, the eyes and the mouth [7].

The research literature in perceptual processing includes face and body detection [8] and automatic extraction of face boundaries [9].

The input measures like colour, brightness, boundaries and motion can be regarded as first-order features in a hierarchical scheme. Such measures are merged to estimate body profiles, eye and mouth movements and location of the facial regions and of the upper limbs.

The next level in the hierarchy consists of a figure parametric description where the motion fields of the eye and mouth regions are computed. In general, such parameter vectors are recorded over time to allow the system to learn some numerical indices of the emotional status. A suitable learning algorithm for this purpose is the learning vector quantization (LVQ).

Finally, the shape and motion of the upper part of the body are detected. As an example, it is possible to evaluate the 3D position of the head and the shoulders along with their movements.

Status information is related to the eye gaze, while transitions are related eye movements. Changes in the eye displacements are classified as gazes or movements depending on their intensity and direction.

Behavioural processing is related to key pressing and mouse data. Key pressing involves the choice of a particular key and the pressure time. Mouse data are the pointer coordinates, clicks' strength and frequency, and movements' acceleration. These are the primal features to update the user cognitive model.

One might think to devise some straightforward rules to adapt the interface layout to the users' emotional status (i.e. in case the user gets confused; the layout has to be as simple as possible). However, this approach is unlikely to be effective because the cognitive status of the user is not taken into account. The best interface arrangement depends on the task the user is engaged with. The system can remove correctly some elements from the interface layout only if the users will not use them, provided the nature of the task they are performing.

In general, the emotional status depends on particular elements in the interface that are related to the particular task in progress. It is better to clarify the meaning of a single object at design time than removing several widgets randomly.

The adaptive system interface module performs true adaptation. Emotional and cognitive status of the user includes confusion, fatigue, low concentration and errors. Adaptation can result in interface simplification, critical information highlighting and tutoring about incorrect actions.

If the facial expression and the gaze indicate that the user is not paying attention to the task and the cognitive model interprets this fact as a reduction of cognitive resources (i.e. mental fatigue), it is possible to reduce the overall workload and to highlight the most critical components of the interface.

Possible adaptation techniques are

- insertion and/or removal of operating features,
- insertion and/or removal of help and feedback dialog boxes,
- changes in data formatting and
- insertion and/or removal of simple secondary tasks [10].

3.1 Perceptive processing

This section deals with the tools used in perceptive processing, such as detection of arm's small movements, facial data elaboration, gaze capture and mouse tracking. Moreover, one can develop new tools, i.e. to detect the upper body (head and shoulders) [10].

We can devise static and dynamic features in non-verbal communication. Table 1 reports a classification of such features.

3.1.1 Limbs' movements

This paragraph describes a technique for searching, detecting and interpreting little arm-hand movements starting from colour video sequences. The approach

Table 1: Classification of static and dynamic features in non-verbal communication.

Static features	Dynamic features
Face	Attitude
Physical structure	Posture, gesture and movements
Tone of voice	Expression
Physical contact	Gaze direction
Proximity	Nods
Appearance	Speech fluency

is suitable for input analysis and to recognize hand motion. As a perceptive processing tool, this method is aimed to transform percepts in a sub-symbolic representation, which encodes an emotional status. In turn, this representation provides the cognitive processing module with a suitable input.

At first, the normal flow field is computed from the input sequence. The expectation maximization (EM) algorithm is used to fit a Gaussian to the normal flow histogram computed across the frame. The moving arm is detected as the dominant region in the normal flow field, that is, the set of all the points whose normal flow value is $\geq 4\sigma$. Sample points are selected as the ones with a large gradient value as well as a large normal flow value and whose gradient is similar to their neighbours. The boundary of the arm is obtained using the Dijkstra shortest path connecting all the sample points. Then, affine transform parameters describing the arm global motion are estimated from the analysis of the arm boundaries.

Symbolic information is obtain via a hierarchical clustering. LVQ is used to compress affine parameter vectors and to derive a labelled Voronoi tessellation where each tile corresponds to a motion primitive without a precise meaning. The next layer clusters label sequences in sub-activities like *up*, *down* and *circle*. Finally, a robust matching procedure based on nearest neighbour classification, groups sub-activity sets into complex sequences like *striking*, *pounding*, *swirling* (= repeated *circle*) and so on.

Figures 2–4 show some steps of the approach explained earlier. Figure 2(a) shows an image taken from a 400-frames long pounding sequence, while Figure 2(b) shows an image from a 100-frames long swirling sequence. Images were captured by a progressive colour scan SONY DFW-VL500 camera with a frame rate of 30 frames per second, each frame being 320 × 240 pixel wide.

Figure 3 (first row) shows the points with the maximum normal flow, while the arm boundary is depicted in the second row.

Figure 4 (first row) shows the residual flow that is computed as the difference between the normal motion field given by the affine parameters and the normal flow. Figure 4 (second row) shows re-estimated affine flows after outlier removal using again EM to fit a Gaussian distribution residual flow.

3.1.2 Facial data elaboration

All the animals, and humans above all, use face as the main channel for non-verbal communication. Expression is composed of several features like eye movements, mouth and eyebrow position, configuration of facial muscles and so on. All these signals are part of a perceptive model that is used as the basis for understanding emotions in the user [11].

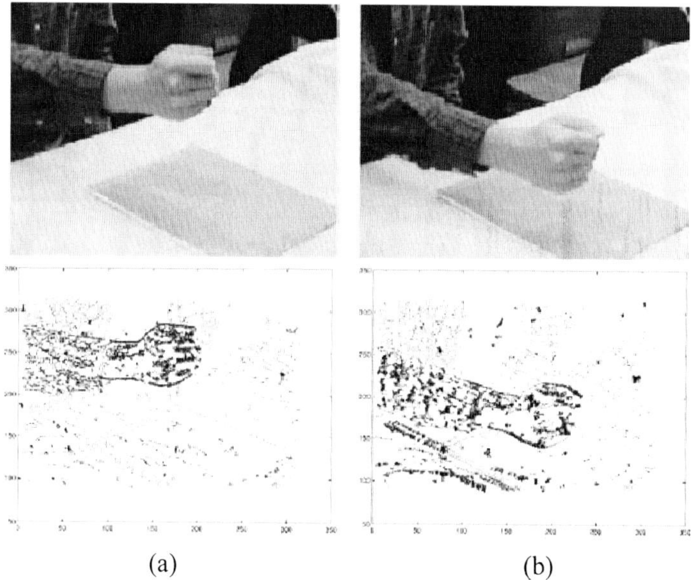

(a) (b)

Figure 2: Two example frames taken from (a) a pounding and (b) a swirling sequence along with the corresponding normal flow (bottom row).

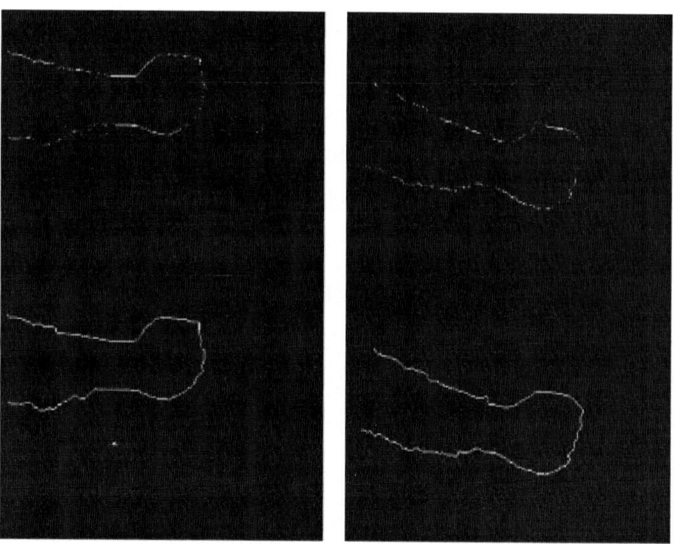

Figure 3: Maximum normal flow points (first row), and tracked contours (second row) for the sequences depicted in Figure 2.

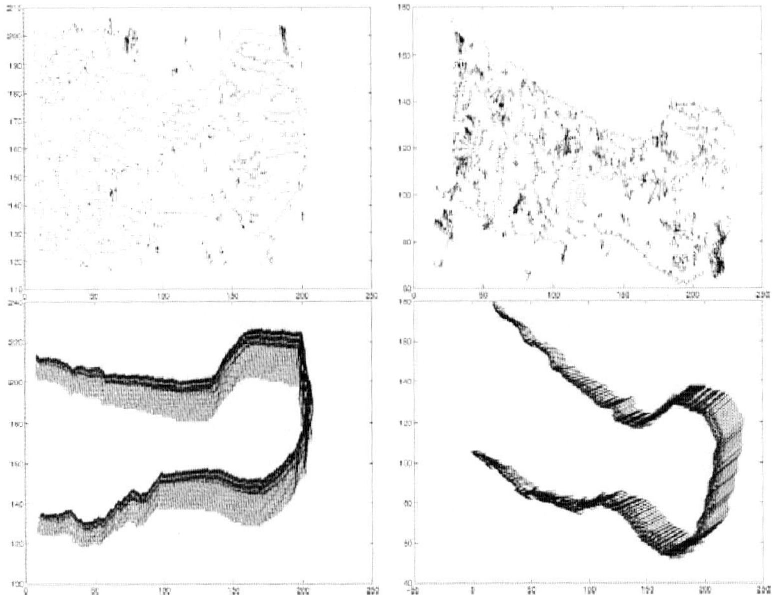

Figure 4: Residual (first row), and re-estimated flow (second row) for the sequences depicted in Figure 2.

As regards the eye, the visual features involved in recognizing the cognitive status are gaze direction, dilation of the pupil and eyelid closure. Figure 5 shows the measures for these parameters.

Dilation of the pupil indicates that cognitive activity is intensifying or it is arising suddenly, while a fixed gaze can be a sign of a very strong mental activity due to data elaboration. Other features are the position of eyelids with respect to iris, the eyebrows' shape (plain, raised or frowned) and the presence of wrinkles at the eye corners, between the eyebrows or beneath lower eyelids.

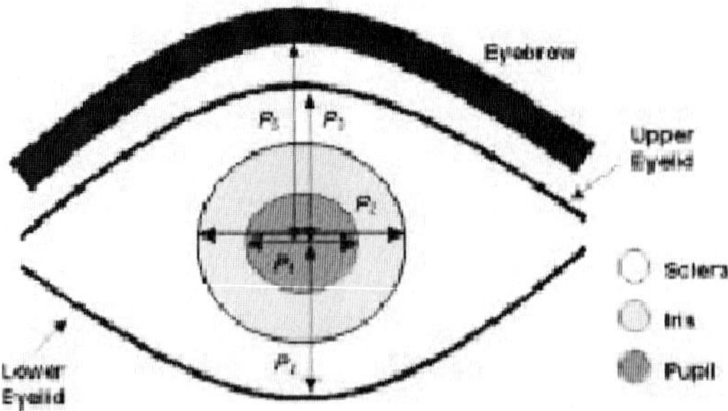

Figure 5: Eye parameters [10].

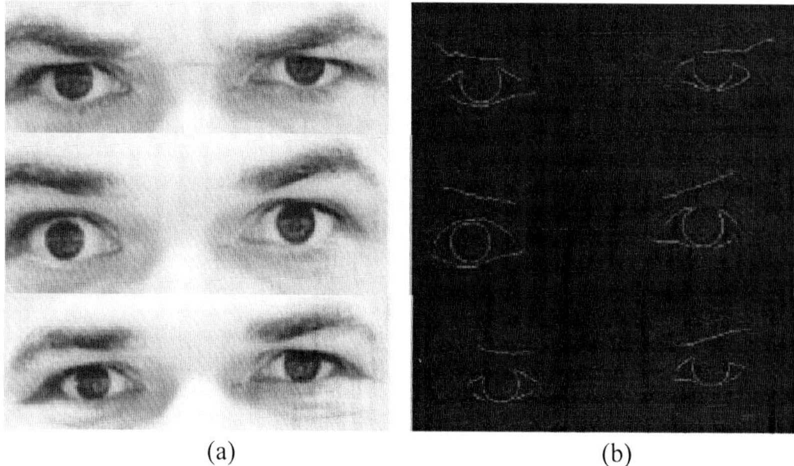

(a) (b)

Figure 6: Visual expressions (a) and the extracted features (b) for three base emotions (one per row) [10].

As an example, completely open eyes where lower eyelids are down, while eyebrows are raised and curved indicate surprise. Eye images should be acquired using high-resolution colour cameras.

Figure 6(a) shows the eye regions corresponding to anger, surprise and happiness. Figure 6(b) shows irises, eyelids and eyebrows extracted from the images on the left side.

3.1.3 Visual perception

Eye-gaze tracking to evaluate the user's cognitive status relies on the immediacy assumption (people elaborate pieces of information while they are looking at them) and on the eye–mind assumption (the eye is fixed on a particular object while it is being analysed). There are several eye-gaze tracking methods. If we focus on data collected from eyes directly, there are two main techniques:

- detection of the corneal reflection from a lit eye and
- iris recognition from eye pictures.

One can choose between these two approaches on the basis of the external lighting conditions. Moreover, there are three methods to compute fixations. The first one implies simple computations because it relies on fixed geometry. In this case, the users have to hold their head still using a restraint. In the second technique, the user wears a head-tracking sensor that is able to detect 3D position and attitude of the head and merges them with the data about the eye direction. The third technique uses an eye-tracking device and a camera both placed on the head and provides an image of what the user is looking at.

The first method is the most accurate but cannot be used in practical HCI. In the same way, head-mounted equipments are not a practical choice. In general, remote camera eye tracking is performed despite of its low accuracy.

The most common method to devise if the user is attentive is to define a distance/time threshold: when two look-at points are close more than a threshold for a sufficient amount of time, a fixation is detected.

Salvucci and Anderson [12] developed a more sophisticated technique that classifies the eye movements using a Hidden Markov Model (HMM). At first, a two-state HMM is used to separate fixations from saccades. These are very noisy data, so a second HMM is used that takes into account the closeness of each fixation to the screen objects and the context made by the other objects the user has just fixated. The model is then compared with several plausible sequences, and the most likely one is selected (best overall fit). Fixations carry information about their position and duration. Position indicates the objects the user has probably dealt with. Duration indicates the objects the user has most likely involved in detailed computations [10].

3.1.4 Voice perception

Voice perception implies redundancy removal from the sound wave, and an effective representation of the main speech features to simplify successive computations. One of the main applications in the field of speech processing is digital encoding of voice signal for efficient storing and transmission.

Vocal communication between humans and computer consists of two phases:

- text-to-speech (TTS) and
- automatic speech recognition (ASR).

Obviously TTS is simpler than ASR due to the asymmetries in producing and recognizing speech.

Two main processes are crucial for both ASR and TTS systems:

- segmentation and
- adaptation.

Segmentation has to be faced both by TTS and ASR. In the case of ASR, segmentation can be helped by particular speech styles. Fluent speech recognition allows the user to have a natural dialogue with the system, but it is a very hard task.

As regards adaptation, users are inured to harsh synthetic voice of TTS systems, while ASR ones have to adapt to any voice. Nowadays, ASR systems require the users to modify their speech using pauses and speaking slowly [11].

3.2 Behavioural processing

Behavioural processing is focused on two data input modalities: keyboard and mouse. Besides being used to have direct interaction with the system, these devices can provide useful information to know the user's cognitive status. Such knowledge is built by tracing both key pressing and mouse positions or actions.

One can devise two main categories in mouse motion data: gestures and clicks. Gestures are mouse motions that do not give rise to clicks, like giving the focus to a GUI element. In general, they are not related to particular functions but they can provide information about the objects the user is processing even if no explicit action is performed.

Three main features define all clicks: velocity, strength of the click and motion readiness towards clicked objects. All these features indicate the degree of excitement, indecision and confusion.

3.3 Cognitive comprehension

This module consists of a cognitive model and a model tracing function. A cognitive model is intended to carry on tasks in a similar way as humans do.

The module builds a detailed map of the interpreted motion data in terms of couples {motion, emotional state}.

Integration of sub-symbolic emotion parameters in a cognitive model is a very important scientific result that can be applied to different operational contexts (i.e. simulations of real-time systems) to trace changes in human performance over time. Cognitive models could be very important tools for the designers of such systems. The tracing process can be helped a lot by mouse gestures, eye data and information about the emotional status.

Basically, emotions can be integrated in a cognitive model in three ways. Emotions can be regarded as modifiers of the model's parameters, while producing simple changes in behaviour. As an example, fatigue can diminish human elaboration speed that is thinking readiness. In the same way, confusion or being sad can be regarded as a noise parameter in the decision process.

Emotions can influence the core structure in the cognitive model. People who are fatigued, sad or confused can completely change the way of thinking about the task they are involved and use different strategies to carry them on.

Finally, one can devise a hybrid framework mixing both the previous approaches.

4 Input/output devices

Selecting the right input and output devices is a crucial point to obtain simple and efficient interaction. Such devices can be modified and/or adapted for being employed by particular users like the disabled. Moreover new devices can be designed for this purpose.

4.1 Input devices

Interaction can start when the user is enabled to communicate her intentions to the machine so that they can be processed. An input device is able to transform the information provided by the user into data that can be recognized and understood by the system [11].

When selecting the role of a particular input device, the interaction designer has to keep in mind that it will help users to carry on their work in a secure, efficient and pleasant way.

In general, input devices should exhibit the following features:

a) They are designed for adapting to the user's physiological and psychological features; this is a very useful feature to obtain devices to be used by disabled or unskilled people.
b) They are suitable for the tasks involved in the interaction; as an example, to draw lines or sketches a tool like the electronic pen is needed that allows continuous movement over a surface.
c) They are suitable tools in the environmental context of the interaction; as an example, a vocal interface could be useful to avoid the keyboard but only when the environment is not noisy [11].

4.1.1 Keyboards

A keyboard is a set of two-state buttons that can be pressed (on state) or not (off state: default). The user can press down single keys or a combination of them. Keyboard is a discrete device because it does not allow continuous interaction [11].

4.1.2 Pointing devices

Pointing devices are used when a point or a line have to be pointed out or selected in the 2D/3D space. Some of them are joystick, trackball, mouse, electronic pen and so on. Pointing devices allow continuous interaction even if mouse and joystick allow discrete interaction through their buttons.

Joysticks act along two directions, and they are sued often when the task involves specifying a direction or speed value. Joysticks can be used in plants when a ubiquitous system is used to move an equipment remotely [11].

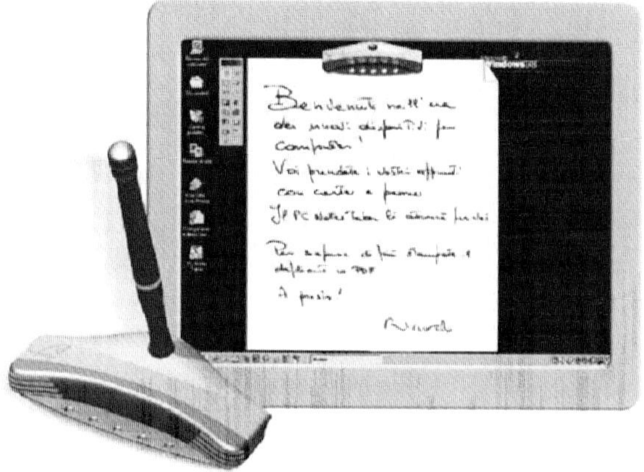

Figure 7: The electronic pen.

A trackball is a small ball that is fixed to a support and can be moved along any direction. A mouse uses a similar principle but the ball allows it to be moved on a flat surface. The electronic pen (Figure 7) is used to digitize whatever the user draws. It is coupled with a tablet: when the pen is passed on the tablet its trace is recorded and digitized as an image. The main drawback associated with this device is writing recognition that has to be performed using dedicated software [11,13].

The touch screen is a pressure sensible screen without movable parts. It can be used by unskilled people too and has the advantage that keyboards can be simulated inside the screen thus reducing the working space [11].

The eye-tracker uses infrared sensors to acquire eye movements. The users can control the pointer on the screen simply by fixating their gaze on it. Nowadays this device has low accuracy.

4.1.3 Cameras

Cameras are a very effective interaction system because video streams can be processed nearly in real time to extract implicit inputs [10]. A UC framework should be equipped with one or more cameras with dedicated computing resources to increase the rate of processed images per second [10,11].

4.1.4 Audio input devices

Audio data need very less storage space than video. Microphones are present in all the vocal interfaces that are very useful for disabled people. Many learning algorithms have been already developed for real-time command interpretation. All of them need a training phase to recognize the user's voice.

4.2 3D input devices

Virtual reality is a well-suited technology for UC. It requires particular input devices like the data glove and the tracker.

The data glove is a real glove equipped with many sensors. It owns accelerometers and gyroscopes to obtain the whole attitude and speed of the hand and motion sensors to detect the position of each finger joint with respect to the palm. A data glove can be used also in a classic 3D graphic environment where the user can manipulate 3D objects depicted on the screen. In general, it allows a very easy interaction but is cumbersome and expensive [10,11].

The tracker is also called spatial mouse. It has the same functionalities of the data glove but the users hold it in their hand [11].

In virtual reality, both data glove and tracker are used with a head-mounted display that can be realized as a helmet or as particular glasses displaying a stereo couple of each image taken in the virtual environment.

4.3 Output devices

A full integration of UC in everyday life needs output modalities that differ severely from traditional ones. The computer screen output will no longer be the only way the machine can communicate with us. In the near future, aural, visual and haptic outputs will be disseminated all around to provide us with effective communication [13].

Output devices are all peripherals and channels that the system uses to communicate with humans [11].

Output selection is crucial for a UC system to work properly. Multimodal communication using visual, aural and haptic channels can be used to simplify the message from the system to the user [11].

One can think to generate 3D output both as 3D graphics and as virtual reality. Spoken messages can be more straightforward than visual ones; the problem remains about the possibility for the system to have sufficient speech samples both as natural and synthesized voice [11].

Some experiments have been conducted about haptic output to be used for special users like the disabled [14].

4.3.1 Visual outputs

Visual outputs are needed to provide whatever user needs with clear and understandable information. The system has to give feedback about the interaction without boring the user. Also the particular widgets that are present in a GUI at a certain moment can be regarded as visual output along with their layout.

Each device in the ubiquitous framework has to be able to give feedbacks about the elaboration state, to alert the user to input data, acknowledge the user for received input or tell the user that the input is wrong [10].

4.3.2 Aural outputs

In current interfaces, aural perception is integrated with visual stimuli to make easier the dialogue between man and machine [11].

Different sounds are used to alert the user on inserted wrong data or about an ongoing computation and so on. In general, sound enriches the information exchanged through the interface [10,15].

Aural perception is crucial in all the application where the user attention is not focused on the screen. Some examples are medical applications, flight, transportation and video surveillance where the attention decreases with time. Finally, blind people can navigate structured data as web pages using suitable software called screen readers [11,16].

4.3.3 Haptic outputs

In the last three decades many haptic devices have been developed but only a few are available because they are very expensive [11]. When low-cost haptic devices can be integrated in UC frameworks, they will be very useful above all for seeing-impaired people [14]. Finally, haptic devices can be suitably integrated in virtual reality [11].

5 Usability

5.1 Relevance for ubiquitous computing technologies

Modern computers are no more tools for technicians alone. More and more people use the PC due to lowering prices and software applications that can be used efficiently even by unskilled people. Easy-to-use software relies on GUIs and their usability.

Usability is defined by ISO norm 9241 as 'an important consideration in the design of products because it is concerned with the extent to which the users of products are able to work effectively, efficiently and with satisfaction'.

Designing usable systems has been a crucial topic in software evolution. When technology was not mature, it was very difficult and expensive to produce usable software. In the field of UC, designers have to pay attention to usability because UC technologies can modify the current usability paradigms.

In this respect, the context-aware systems are very interesting. These are software architecture that can perceive the surrounding environment and extract useful information about both their processing tasks and the interaction status. So a new generation of adaptive interfaces is being developed right now.

5.2 Usability issues and technology changes

Usability problems in UC cannot be solved in any situation. GUIs are not always at disposal in UC systems, while interaction cannot be limited to a classical display-keyboard-mouse arrangement. Moreover, there are no more expert users; we are not sure that either the user wants to use particular software or has the required skills to use it. In general, UC frameworks are general-purpose systems. They are intended to satisfy heterogeneous requests from very different people with regard to competency and skills.

Users would not always interact with visual devices or they will be unable to do that. UC systems will make use of different communication modalities like biometry or speech recognition so that the system will interact autonomously with the user, which would not care of being understood by the machine.

Severe modification in both hardware and software systems are needed. AI and user modelling can help the designer on the software side, while new interaction devices are needed on the hardware side that allow the user to achieve natural interaction.

6 Portability

6.1 Why 'information portability'?

At first, one might think that information portability is aimed to complete interoperability between heterogeneous wired and wireless communication systems. The main goal of portability is allowing any user to access their workstation whenever, wherever and using whatever terminal. As a consequence, the users are enabled to use their office automation tools and access enterprise data and the information they are interested in.

Internet represented a first step towards information portability. Wireless devices represent another crucial improvement; they ensure connection regardless the actual position of the user.

Mobile devices do not imply portable information. In general, a mobile terminal is intended for access to particular services and allows limited portability. A true mobile terminal has to provide the user with Internet connection, to be equipped with a minimal application suite and to be able to execute remote applications. In other words, a mobile terminal is required to be an element of a distributed system.

Portability relies on both hardware and software mobility. Hardware devices have to be mobile and have to operate in any environmental condition; on the other side, mobile software can be executed on heterogeneous mobile hardware [17].

6.2 Some issues about portability

Mobile devices are very dissimilar from classical computers. Computers use the screen as their interface and can enrich information through visualization. On the contrary, mobile devices own very small displays that have to visualize a huge amount of information.

Information visualization is one of the most crucial issues related to portability. The same information has to be visualized by very different displays, so the designers have to keep this in mind when they devise visualization techniques. Diversity influences both the structure and the format of information.

At design time, the designer cannot know the features of all computing platforms. The only consideration to keep in mind is that visualization has to be designed for the user and not for the publisher. The final user is the reference for information portability.

One can design for portability according to two approaches. The former is the design completely adaptive applications that change their appearance as a function of the hardware they are running on. This leads to hardware and software independence. The latter technique is to develop a set of discrete levels of adaptation to be selected on the basis of the hardware.

Regardless the design approach, users with different mobile platform will have a different interface for the same application that will be suitable for their hardware. At the same time, each interface will be compliant to the most widely accepted standards [18].

7 Conclusions

An UC system can be defined as a computing system that can be located in any building (house, office and so on) that is able to process and exchange information with other systems. Users inside the building are continuously monitored by a sensor network to capture their movements and their actions.

Even if this is very interesting for computer scientists and engineers, several criticisms come from non-technical people about all the problems deriving from privacy protection. UC will be widely accepted only when suitable technologies for privacy protection will be in action.

UC is not augmented reality; it makes computing resources available in all environments. UC will surpass current ideas about the computer as a basis element of our working space. In UC vision, computers will disappear to be integrated in the environment and in the objects of every-day use [15].

Several new technologies will be developed in the forthcoming years that will be able to extract pieces of information from our movements and actions, to process them, to devise our physical and/or mental status and to decide if we need care. Many theoretical studies about the needs of the people interested in UC were the first step towards this goal. Scientists tried to devise if impaired people can be also among UC users [10].

In general, UC applications have to serve very heterogeneous requests; this led to the design of personal interfaces that are suited to the single user performing a request.

UC technology has been progressively introduced in everyday technological objects, while reusing well-known interfaces for a new generation of applications. This strategy has been adopted to reduce the traumatic effect of the new technology.

Communication between UC devices will make it possible for the users to monitor their house remotely while they are driving or to control kids or elder people.

While using UC systems, one must keep in mind that they are intended for decision support and that they do not replace humans in the decision-making process [10]. All scientific studies related to UC technologies are devoted to build systems that are able to 'think like humans' without replacing the user. Such systems will be able to decide in place of the user when there can be a health emergency or the danger of death.

Acknowledgements

This chapter was written with the contributions of the following students who attended the lessons of 'Grids and Pervasive Systems' at the faculty of Engineering in the University of Palermo, Italy: Bennardo, Bommarito, Cannizzaro, Carlevaro, Carmicio, Castellino, Cefalù, Ciuro, Di Trapani, Failla, Genco, Guglielmini, Iacono, Inguanta, Lo Cascio, Lo Iacono, Marino, Portuesi, Sicilia and Tusa.

Authors would also like to thank Roberto Pirrone for his help in the Italian–English translation.

References

[1] Weiser, M., The computer for the 21st century. *Scientific American*, **265(3)**, pp. 94–104, 1991.

[2] Schmidt, A., Implicit human–computer interaction through context. *Personal Technologies*, **4(2&3)**, pp. 191–199, 2000.

[3] Abowd, G. & Mynatt, E., Charting past, present and future research in UC. *ACM Transactions on Computer–Human Interaction*, **7(1)**, pp. 29–58, 2000, Special issue on human-computer interaction in the new millennium, Part 1.

[4] ISO 9241-11, *Guidance on Usability*, http://www.usabilitynet.org/tools/r_international.htm, retrieved on June 2009.

[5] ISO 9241-11, *Guidance on Usability*, http://www.usabilitynet.org/tools/r_international.htm, retrieved on June 2009.

[6] Gonçalves, D.J.V., UC and AI towards an inclusive society. *WUAUC'01: Proc. 2001 EC/NSF Workshop on Universal Accessibility of Ubiquitous Computing*, ACM Press: Alcácer do Sal, Portugal, pp. 37–40, 2001.

[7] Jabri, S., Duric, Z., Rosenfeld, A. & Wechsler, H., Detection and location of people in video images using adaptive fusion of colour and edge information. *Proc. 15th Int. Conf. Pattern Recognition (ICPR'00)*, September 3–8, Washington DC, Vol. 4, p. 4627, 2000.

[8] Huang, J., Gutta, S. & Wechsler, H., Detection of human faces using decision trees. *Proc. Int. Conf. Automatic Face and Gesture Recognition*, Killington, VT, pp. 248–252, 1996.

[9] Sirohey, S., Rosenfeld, A. & Duric, Z., A method of detecting and tracking irises and eyelids in video. *Pattern Recognition*, **35**, pp. 1389–1401, 2002.

[10] Duric, Z., Gray, W.D., Heishman, R., Li, F., Rosenfeld, A., Shoelles, M.J., Schunn, C. & Wechsler, H., Integrating perceptual and cognitive modelling for adaptive and intelligent human–computer interaction. *Proceedings of IEEE*, **90(7)**, pp. 1272–1289, 2002.

[11] Jaimes, A. & Sebe, N., Multimodal human–computer interaction: a survey. *Computer Vision and Image Understanding*, **108(1–2)**, pp. 116–134, 2007, Special Issue on Vision for Human–Computer Interaction, DOI:10.1016/j.cviu.2006.10.019.

[12] Salvucci, D.D. & Anderson, J.R., Automated eye-movement protocol analysis. *Human–Computer Interaction*, **16**, pp. 39–86, 2001.

[13] Abowd, G.D., Mynatt, E. & Rodden, T., The human experience [of ubiquitous computing]. *IEEE Pervasive Computing*, **1(1)**, pp. 48–57, 2002, DOI: 10.1109/MPRV.2002.993144.

[14] Lee, K. & Kwon, D.-S., Sensors and actuators of wearable haptic master device for the disabled. *Proceedings of International Conference on Intelligent Robots and Systems. IEEE/RSJ*, **1**, pp. 371–376, 2000, DOI: 10.1109/IROS.2000.894633.

[15] Rehman, K., Stajano, F. & Coulouris, G., Interfacing with the invisible computer. *Proceedings of the Second Nordic Conference on Human–Computer Interaction*, **31**, pp. 213–216, 2002.

[16] Frances, T.M. & Janice, R., *Guidelines for Accessible and Usable Web Sites: Observing Users Who Work with Screen Readers*, self-published version. Redish & Associates, 2003, http://www.redish.net/content/papers/interactions.html, retrieved on July 2008.

[17] Kumar, P. & Tassiulas, L., Mobile multi-user ATM platforms: architectures, design issues, and challenges. *IEEE Network*, **14(2)**, pp. 42–50, 2000, DOI: 10.1109/65.826371.

[18] Nielsen, J., *Designing Web Usability: The Practice of Simplicity*, New Riders Publishing, Thousand Oaks, CA, 1999, ISBN 1-56205-810-X.

Chapter 4

Disappearing hardware

1 Introduction

Although the physical world is far from being a single, interactive computing platform as many experts believed, it is under everybody's eyes that we are more and more dependent on computer systems, nowadays embedded in a growing number of everyday life objects, often already connected in networks. It is also true that, today, using a computer is still a difficult task for many, where much of the effort is spent handling the complex dialogue of interacting with the inanimate box rather than on the task itself, which compelled us to use the computer in the first place.

Currently, personal computers are much more than an inanimate box, but they still require knowledge, experience and effort for their correct use. Ideally, computer systems should infer a users' intention by a small set of gestures or commands and then, as much as possible, perform the task autonomously. Much of the research herein discussed is aimed at designing a new paradigm for computer systems, in which computing capabilities are embedded in everyday life physical objects. This concept, now enabled by the current technology advances, was already introduced in 1990s by Weiser, who termed it *Ubiquitous Computing* in his seminal work '*The Computer for the Twenty-first Century*' [1]. Instead of the traditional *point-and-click* interface, these systems will enable a more natural interaction depending on where they are. The object should be, in other words, aware of the surrounding environment and be capable to sense other objects around it, communicating with them and keeping memory of its own past activity [2]. These computing objects will then be capable to operate depending on the context, showing a sort of *intelligent behaviour*.

The goal then becomes to design computer systems that are easy to use and do not distract users from the task they intend to accomplish by using them. In other words, computing resources must become *invisibly part of our daily routine* [3] or belong to *walk-up-and-use* systems [4] that you can use with knowledge limited to the task at hand alone.

The rest of this chapter is organized as follows. Section 2 addresses invisibility as a key paradigm for ubiquitous systems. Section 3 is focussed on evolving hardware. Key issues for building of ubiquitous systems are covered in Section 4. In Section 5 a quick overview of proactive systems is given. A discussion on problems and limits of invisibility is given in Section 6. Conclusions and relevant references then close the chapter.

2 Invisibility, a key paradigm for ubiquitous systems

Ubiquitous systems are strictly related to the concept of *invisible computer*. Weiser, in fact, suggests that better technologies are those in which the physicity of the technology remains in the background, leaving users free to act, under the impression that they are performing a task rather than using a tool. A tool or device is thought to be invisible when it is integrated in the environment to the point that users can interact with it without even noticing that they are using it. Such artefacts will constantly lie in the back of users' attention.

A personal computer is, as we currently know it, a *primary* artefact, i.e. an object that is normally perceived as extraneous to the surrounding environment. When computing devices are embedded into everyday life objects they become *secondary* artefacts, ideally invisible and capable to interact implicitly with other secondary devices. Secondary devices augment an object's capability to process and exchange digital information between them and with users; however, they should leave unchanged the semantic meaning of the object in which they are embedded, maintaining their properties, use and physiognomy. This way, computer disappear as perceptible devices, while augmented objects emerge [5].

We will witness then to the development and spreading of minuscule computing devices, wirelessly interconnected, hidden inside common objects to the point of being invisible to the human eye. This trend is already anticipated by many specialized devices already available in the market, such as automated language translators, web pads, e-books and so on, in which computing resources and interface capabilities are strictly related to the specific purpose for particular object at hand. Generally, a person is compelled to interact with a computer mostly because of the need to access digital information or to collaborate with other people, rather than being interested to the device for itself [6,7]. Evidently, invisibility is not to be referred to physical dimensions of a component but rather to its seamless integration with the surrounding environment. Weiser in [8] states that a good instrument is an invisible one, citing as example the eyeglasses. While wearing them, a person does not focus on them as a primary object but rather on seeing better. This is definitely not the case with a personal computer, which still forces the user's attention on the instrument itself rather than on the task to be accomplished, so often captured by stalls and failures.

Dimensions are thus irrelevant, and Weiser states that invisible systems may be built at all scales, inch-, foot- and yard-scale [1] as long as users are not distracted by the devices in which they are immersed [9].

2.1 *User-centric* versus *desktop-centric* systems

Let us consider a current desktop system: it is intrinsically devoted to a single user who may perform a single or multiple concurrent tasks. For instance, user may be writing an email or reading a web page, while listening to a music track and scanning the hard drive for viruses and trojans. In all cases, the running computer programs are in control of the interaction, leaving the user to choose within a predefined set of possibilities; the interface may change its aspect, at times presenting a command line while a windows environment in others, but the users must always be knowledgeable of the interface syntax at all times and renounce to their freedom of choice in how to accomplish the task. As this interaction occurs mostly at *machine level*, users are subjected to the following three unfortunate consequences: (i) must be familiar with the computing device and well-trained on using its interface; (ii) must create their own mental map to associate labels and icons to specific tasks and devices and (iii) perceive the interface as unnatural and get easily distracted from the task.

To this *desktop-centric* vision, the ubiquitous systems community opposes a *user-centric* one, in which the user is at the centre of the environment and computing systems are embedded all around to support the needs and desires. In other words, users move freely in their selected environment, be their house, the office or a public space and the objects in the environment sense their presence and act according to the users' requests. The point of view is now reversed, as the flow of interaction is now controlled by the individual and not by the computer, Computers become *context-aware*, as it is conscious of its capabilities and owns information about the environment where it operates. These computer devices can both relate to each other and to the users, dynamically adapting to the needs of the user controlling them [10]. Interaction, therefore, moves from an explicit level to an implicit one.

2.2 Environment-distributed systems

As mentioned earlier, dynamic adaptability of systems to ever-changing user needs is achievable by setting aside the traditional interaction modality: fix workstations disappear and mobile devices gain renewed importance. Computing systems are distributed into the environment and communicate by short- or long-range wireless networks. Differently from their desktop counterparts, these computing devices are designed as *task-specific*, with stringent limitations to their energy consumption so that memory and processing capacity are equal to what is needed to accomplish the task and not more. For these systems, the interface itself serves to communicate directly to the user the capabilities of the system and its current state of cooperation with neighbouring devices; no need for a special training, but rather a means that let the user intuitively infer what the device can do for him. The interface is no longer a shell around the computing system inside but rather a layer transforming user actions into commands for the underlying device, without exclusively controlling the dialogue with the user.

Now that computing devices are distributed into the environment and each of them is dedicated to a specific task, how is the user going to interact with them? Weiser suggests in [1] to use '*many many displays*', maybe reusing those already available around, such as TV screens or picture frames. Touchscreens and voice interaction may serve as input devices. In all cases, input and output terminals are uncoupled from the distributed computer systems and connect to it via wireless links.

The fundamental purpose for these new systems is to make the computer a mere instrument, no longer being the focus of the interaction. It is embedded as a secondary artefact in many everyday life objects, which maintains aspect, purpose and functionality according to their common use, augmenting them for processing and digital information exchange capabilities [5].

3 Evolving hardware

Initially conceived to replace man in repetitive calculations, computers have evolved into real assistants for many human tasks. The human–computer symbiosis envisioned by Licklider in 1960 [11] was a radical intuition at that time and it took 20 years to become the real purpose behind using a personal computer (PC). From instruments useful for many tasks, computers are today a means of communication, thanks to the tremendous spreading of World Wide Web and its technologies (Figure 1).

Figure 1: Today's computers: attention is devoted to a restricted number of devices.

The development of ubiquitous systems will induce a progressive disappearance of hardware as we know it today, distributing it on objects and environment so that we will then think of a computer as an accessory of everyday life objects, an ingredient of a complex recipe rather than the recipe itself.

Weiser's vision of ubiquitous systems and computing has been embraced by the research community, and the first attempts maybe conducted in the studies at the Xerox Research Centre in Palo Alto (Xerox PARC) in the early 1990s, with their best-known projects as ParcTab, Mtab and LiveBoard. ActiveBadge by Olivetti Research and InfoPad by UC-Berkeley also embraced the same line of research, as other research centres did, such as Carnegie Mellon University, Georgia Institute of Technology, Rutgers University, Washington University and IBM.

Technology was less than ready at the time, so these devices could not fulfil designers' expectations. No wireless connectivity was available, screen capabilities were rather limited (Figure 2(a)), embedded processor were running at a few MHz while their PC counterparts were already speeding beyond 50 MHz, memory was limited to a few hundreds kilobyte versus the tens of megabyte available in PC-mounted hard drive and more.

In less than a decade a number of tablet-shaped products were developed (such as the first IBM ThinkPad, the Apple Newton, the Casio Zoomer and the General Magic Pad), in an attempt to emulate the pen-and-paper approach. Again, hardware was the predominant factor determining the (non-) acceptance of the device, and the final cost–performance ratio did not make

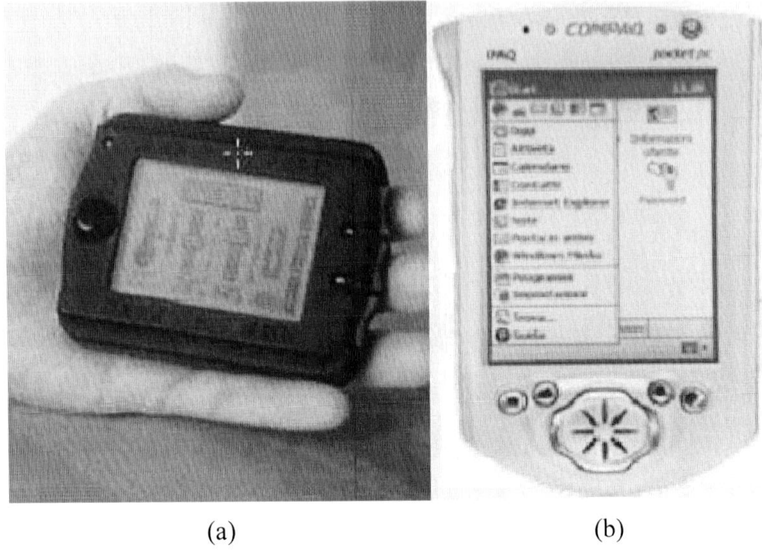

(a) (b)

Figure 2: (a) Xerox ParcTab (1992) first context-sensitive computer, sporting a monochromatic display 128x64 pixel and (b) a current PDA, featuring a colour VGA (240x320 pixel) display.

the mark. To see the first above-threshold product we needed another decade, with the advent of the first families of Palm Pilot, small, light and focussed on a few essential tasks (planner, organizer, contacts manager and a few more) with a single one-button approach to stay in synch with your desktop applications. For the first time, the computer industry produced a device better at one task of the traditional approach, the paper-based daily planners.

To make hardware disappear, its use must be transparent to their user. A slow Wi-Fi connection will end up absorbing our attention away from our task at hand, as much as low-resolution display renders poorly the perception of the entire application. In the very last decade, advances in four technology segments have directly impacted on the development of ubiquitous systems: wireless networks, increased computing power, increased memory capacity, and high-resolution displays. The following sections will briefly discuss them and their relevance to ubiquitous systems.

3.1 Wireless networks

Although slowly at the beginning, wireless networks have advanced at a steady pace, thanks to the development of both short-range (Bluetooth and Infrared Data Association (IrDA)) and long-range (HomeRF, Wi-Fi e Wi-Max) technologies. Today telecommunication infrastructures allows a seamless transition from wired to wireless connections, allowing for ubiquitous connectivity at all levels and distances.

3.1.1 Bluetooth

Bluetooth was developed as a short-range connectivity standard to replace wired networks of small dimensions. It allows successive interactions more efficiently and it allows for localization in the area of limited extension. It consumes a very small amount of energy, which makes it fit the stringent energy requirements of mobile devices. Dedicated USB dongles may be employed to enable Bluetooth in devices that do not have it natively. Its operating range is of about 10 m at 1 mW of power. Bluetooth devices transmit on ISM (industrial, scientifical and medical) band, in the range 2.4–2.48 GHz, with an effective bandwidth of 79 MHz.

It uses spread spectrum transmission to reduce interference with other devices operating on the same frequencies. Specifically, Bluetooth uses the frequency hopping spread spectrum (FHSS) that divides the available bandwidth into 79 bands of 1 MHz each and then define a pseudorandom hopping scheme among the 79 channels that is known only to the devices involved in the communications, thus improving the resiliency to interference.

3.1.2 IrDA

The IrDA pretty much attempts to address the same applications as Bluetooth, simple and quick communications between peripherals. To the purpose of enabling stable communications between mobile devices, however, this technology fails as it requires that the two communicating devices be in *line-of-sight*. This major requirement is due to the fact that it employs infrared light to carry the communication, which is blocked by solid, non-transparent objects. Other limitations are the very short range (below 1 m) and view angle (less than 30°). IrDA devices are typically found on cellular phones, some portable PCs, personal digital assistants (PDAs) and printers. IrDA data rates are normally 4 Mbps, which becomes 16 Mbps with the new fast infrared (FIR) standard.

3.1.3 HomeRF

HomeRF is a wireless technology developed for domestic uses by the HomeRF Working Group. It operates on the same frequency as Bluetooth (2.4 GHz) and employs the Shared Wireless Access Protocol (SWAP), originally capable of 2 Mbps and then extended to 10 Mbps (SWAP2.0). As for Bluetooth, it uses the FHSS technology over six digital enhanced cordless telephone (DECT) voice channels and one data channel following the IEEE 802.11 wireless standard. It does not require dedicated access points as for Wi-Fi, with individual devices connected point-to-point. Major limitation is that its range is limited to 20–40 m and it is difficult to integrate with pre-existent wired networks.

3.1.4 Wi-Fi

Wireless local area networks (LAN) are one of the most relevant access network technologies available today. Their pervasive diffusion in homes, workplaces, universities and schools, cafés, airports and public areas makes them the most readily available way to connect ubiquitous systems among them.

Among the many wireless LAN technologies developed in 1990s, the one standard that has emerged is the IEEE 802.11, also know as Wi-Fi (for Wireless Fidelity). A logo has been introduced by the Wi-Fi Alliance to certify the full compatibility of devices conforming to one or more of the Wi-Fi standards, in a way ensuring interoperability across brands and vendors. Three main standards are available within the 802.11 standard, termed with suffix letters a, b and g. They differs in frequency range and data rate, with 802.11g being the best trade-off between 802.11a and 802.11b, working on a 2.4–2.485 GHz frequency range with a data rate of up to 54 Mbps. Differently from the Bluetooth standard, the Wi-Fi uses a direct sequence spread spectrum method, which offer higher resiliency to errors.

The architecture is based on a *basic service set* composed of a number of *access points* (or base-stations) wire connected to the land network. Each mobile device is then provided with a wireless transceiver that connects to the access point in range through one of the Wi-Fi protocols. A typical coverage is in the range of 150–300 m, depending on the presence of obstacles such as walls, trees or barriers in general.

3.2 Increasing computing power

For the past three decades, Gordon Moore's empirical observation on chip density doubling every 18 months has proven true, with a remarkable increase in computational power along with decreasing silicon footprint and power consumption. This has yielded to the development of chipsets dedicated to mobile applications that are computationally powerful and power savvy, enabling many new applications simply unthinkable a few years ago. Increasing computational power and battery lifetime are, of course, key enablers for the development of disappearing systems.

3.3 Increasing memory capacity

Another key enabler is the increasing memory capacity that new gigascale integration techniques (sub-nanometer transistor channel size) are providing. This is more and more enabling the idea of endless memory capacity even for mobile devices, which can rely on large readily accessible datasets on user's history of interaction, enabling clever behaviours in ubiquitous applications.

3.4 High-resolution displays

As vision is our main sense, disappearing devices must rely on sharp and rich visualization displays. If information is to be shown on low-quality displays, most of our attention will focus on recovering contents from its poor representation. Latest flat screen technology offers today a wide array of visualization devices, some of which combined with precise gesture-based interfaces.

Another rapidly developing technology with a large application potential to disappearing devices is based on organic light-emitting devices (OLED). OLED displays will enable brilliant, low-power, low-cost and flexible displays that can be moulded around the most disparate objects, from textiles, to toys, to home furniture and fixtures, walls and surfaces of all kinds and shapes. Thin-film displays that can be wrapped around things or transparent ones to lay over windows or even printed over T-shirts are new application scenarios.

Plastic electronics based on conducting polymers will enable, in general, a whole new set of devices that can be seamlessly embedded on everyday objects. In 2000, the 1970's discovery of conducting polymers earned a Nobel Prize for chemistry to Alan J. Heeger, Alan G. MacDiarmind and Hideki Shirakawa.

4 Building ubiquitous systems

Two main approaches are followed to build ubiquitous systems: (i) *infrastructure-based systems* that are associated to a particular physical environment and (ii) *personal systems* that are instead user-centric and also include *mobile and wearable systems*. In both cases, new interaction modalities that include dialogue handling and gesture control are a need for enabling direct and natural interaction with the users [12]. In the following two subsections, details will be given on the two approaches discussed.

4.1 Infrastructure-based systems

Developed by Intel, a Personal Server [13] is a mobile device that uses existing computing infrastructures such as PDAs or smartphones to interact with a user's personal data; it does not show information on a display of its own, but rather relies on wireless connection to either large, wall-mounted displays or screens and keyboards of neighbouring PCs that are readily available in the environment.

In this area, a remote display environment is proposed by researchers of the University of Osaka [14] to deploy large, wall-mounted displays in public areas that interacts as remote devices with cellular phones and PDAs.

4.2 Personal systems

On the contrary, personal systems are those that aim at providing an infrastructure that moves with the user himself. Wearable computers are one way to build such systems; however, the common perception of them built of a head-up display, one-handed keyboards and belt-attached PCs does not exactly fit the idea of a disappearing, unobtrusive system. Instead, a sensor infrastructure that is embedded on a man's coat is a better example, although computing capabilities of such systems is still rather limited. *MIThril* is a prototype of such systems [15]. Developed by MIT researchers, MIThril is a *context-aware*, networked set of sensors (including a thermometer for body temperature, heart beat and pulse sensor and a set of microcameras and microphones) embedded on a coat that interacts with the surrounding environment, exchanging personal information among the people present at any given time.

5 Invisibility: problems and limits

As with every other technology that involves a paradigm shift in our every-day habits, practical implementation of ubiquitous systems is still hindered by a set of problems and limits. The following subsections will attempt to highlight the most limiting ones.

5.1 Size and power consumption

Technology advances on both microprocessor size and performances and memory capacity are both of extreme importance to build ubiquitous disappearing systems. While for a PC it would be intolerable that one application grinds to a halt because of another one running on the background, this is not an issue for single task devoted systems as in the case of ubiquitous ones. However, while it is desirable to make devices smaller and smaller as it would be easier to hide them in the environment, too small dimensions become a problem from the interaction point of view, and size cannot be decreased beyond a certain point for those components responsible of direct interaction with the user. Would this not be a problem, having many small devices distributed in the environment may soon become unmanageable as it would be difficult to track and handle all of them at once. Power

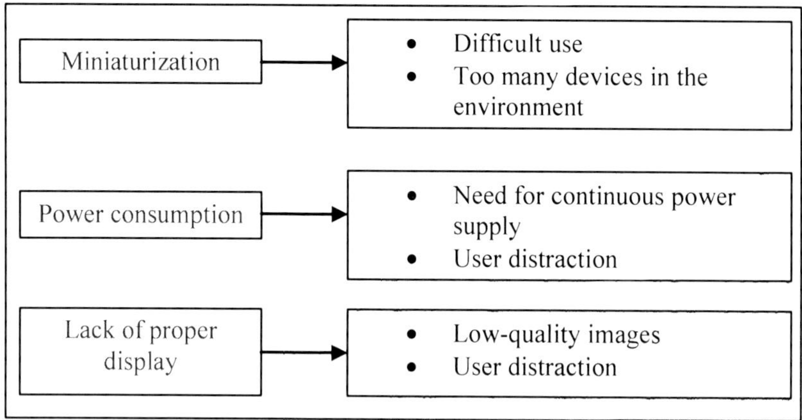

Figure 3: Problems related to size and power consumption.

consumption and the need of providing continuous energy to a large set of devices is then another limiting factor, contrasting with a ubiquitous vision and moving back towards a desktop-centric vision. The following figure summarizes the aforementioned problems and highlights the main issues related to size and power limitations for ubiquitous systems (Figure 3).

Many techniques are employed nowadays on microprocessor to dynamically control individual core energy consumption, mostly connected to software control of transistor operating frequency. Another concurrent approach aims at improving power density for batteries, but also to find alternative ways to derive energy from the external environment, relying on concealed solar cells or piezoelectric sensors capable to extract energy from body movements or body vibrations. Other alternatives look at transmitting energy to the devices, as in *active electronic tagging* [16]. Lastly, display size must always be adequate to the type of contents to be provided and must always provide the appropriate contrast and colour quality to maintain user's attention focused on the contents and not on the device.

5.2 Control loss and lack of feedback

As states Weiser in [1], interaction between users and ubiquitous systems must be dynamic and perceived as most natural as possible. This implies on one hand making fairly autonomous systems capable of deciding by themselves on a number of using scenarios and, on the other hand, depriving the users from the possibility to change system behaviours beyond indicating some basic preferences. This control loss is particularly negative when default behaviours do not match user's intentions, to the point to either distract him or even plainly annoying him. It is thus indispensable to implement *overriding* functionalities

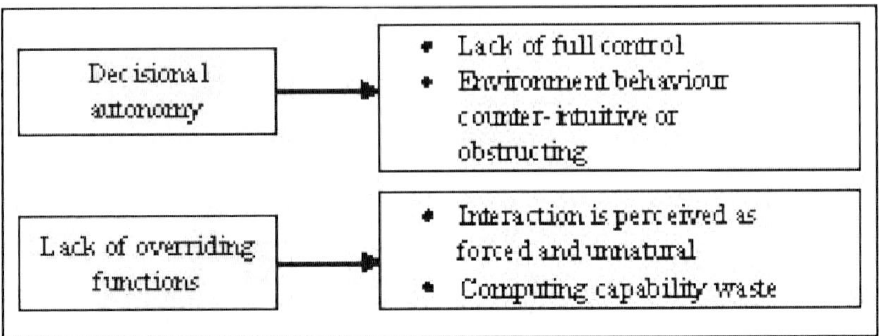

Figure 4: Problems related to size and power consumption.

that enables the users to take control on those circumstances requiring it, but also optimizes power consumption, as in [17].

In all cases, some form of feedback is not only assumed but it must be fully understood by the user. This assumption, however, cannot be held true as the presence of numerous and diverse devices in the environment may make it difficult to recognize which system is operating at any given time and where feedback is to be expected from. In general, it is preferable to embed feedback and control capabilities in everyday objects [18]. Diagnostics must be embedded into the systems and not based on dedicated displays or consoles; moreover, it should clearly indicate the exact location it is referring to, the systems involved and suggested actions (if possible) to correct the situation. At the same time, the diagnostics systems must also continuously interact with the user, ensuring that everything is working properly [4]. Control loss and lack of feedback problems and issues are summarized in Figure 4.

5.3 Breaking the traditional mental model

The design of a PC-based system cannot abstract from the design of a usable interface that complies with Schneiderman's *golden rules* [19]. Besides these physical criteria, the design must be in harmony with the user's *mental model* of overall system behaviour [18,20]: the user must know and somehow expect device capabilities and limitations. These arguments, however, are completely extraneous to the ubiquitous computing paradigm, where a central PC does not exist and many interacting devices leave often the user without choice but that available for the specific interaction context. In a ubiquitous system, there is the need to build a new mental map of the devices at work without any explicit information on their possible uses. Rather, a user must rely on the natural way to evolve with own knowledge, based on two complementary modes: analogies between new and known already experienced situations and stimuli derived from the surrounding environment.

Ubiquitous systems come with a drastic cut with the traditional mental model, in which devices and tasks are no longer structured and prearranged but rather dynamically adapt to changing user contexts [17].

6 Conclusions

Many of the hardware component needed to make Weiser's vision of Ubiquitous Computing true are nowadays available and, as seen in this chapter, can be combined in disappearing systems. Thanks to many technology advances, it is less probable that a user be distracted by the underlying presence of electronics in everyday objects.

Rather, it may still be derailed by the inefficiencies of the interface layer that it is still far from disappearing. Pretty much we need to get to the point that hardware and connected software will become as invisible as is ink on a page: we naturally read it to learn its contents rather than focusing on the ink that makes the words visible on the page.

Acknowledgements

This chapter was written with the contribution of the following students who attended the lessons of 'Grids and Pervasive Systems' at the faculty of Engineering in the University of Palermo, Italy: Arena, Buongiorno, De Vincenzi, Ferraro, Giacalone, Inglisa, La Malfa, Pellegrino, Rotolo and Titone.

Authors would also like to thank Antonio Gentile for his help in the Italian–English translation.

References

[1] Weiser, M., The Computer for the 21st Century. *Scientific American*, **265(3),** pp. 94–104, 1991.

[2] Steven, C.W., *Making Everyday Life Easier using Dense Sensor Networks*, Intel Architecture Labs, Intel Corporation, Hillsboro, USA, 2001.

[3] Tolmie, P., *Unremarkable Computing*, Xerox Research Centre Europe, Cambridge Laboratory, Cambridge, UK, 2002.

[4] Cooperstock Jeremy, R., *Making the User Interface Disappear: The Reactive Room*, University of Toronto, Toronto, Ontario, Canada, 1996.

[5] Michael, B., *Mediacups: Experience with Design and Use of Computer-Augmented Everyday Artefacts*, Telecooperation Office, University of Karlsruhe, Germany, 2001.

[6] Streitz, N., *The Role of Ubiquitous Computing and the Disappearing Computer for CSCW*, German National Research Center for Information Technology, Germany, 2001.
[7] McCarthy, J.F. & Anagnost, T.D., *MusicFX: An Arbiter of Group Preferences for Computer Supported Collaborative Workouts*, Center for Strategic Technology Research Accenture, USA, 1998.
[8] Weiser, M., The world is not a desktop, Perspectives article for ACM Interactions, November 1993.
[9] McCarthy, J.F., *UniCast, OutCast & GroupCast: Three Steps Toward Ubiquitous, Peripheral Displays*, Accenture Technology Labs, USA, 2001.
[10] Pham, T.-L., *A Situated Computing Framework for Mobile and Ubiquitous Multimedia Access using Small Screen and Composite Devices*, Multimedia/Video Department Siemens Corporate Research, Inc., Princeton, NJ, 2000.
[11] Licklider, J.C.R., Man-computer symbiosis. *IRE Transactions on Human Factors in Electronics*, **HFE-1**, pp. 4–11, 1960.
[12] Kangas, K.J. & Roning, J., *Using Mobile Code to Create Ubiquitous Augmented Reality*, University of Oulu, Department of EE, Computer Engineering Laboratory, Finland, 2002.
[13] Want, R., *The Personal Server: Changing the Way We Thing About Ubiquitous Computing*, Intel Research, Santa Clara, CA, 2002.
[14] Uemukai, T., *A Remote Display Environment: An Integration of Mobile and Ubiquitous Computing Environment*, Department of Information Systems Engineering, Graduate School of Engineering, Osaka University, Japan, 2002.
[15] DeVaul, R.W., Pentland, A. & Corey, V.R., *The memory glasses: subliminal vs. overt memory support with imperfect information.* Seventh IEEE International Symposium on Wearable Computers (ISWC'03), White Plains: New York, October 21–23, p. 146, 2003.
[16] Borriello, G. & Want, R., *Disappearing Hardware*, University of Washington and Intel Research, Seattle, WA, 2002.
[17] Rehman, K., *Interfacing with the Invisible Computer*, Laboratory for Communications Engineering, Cambridge University Engineering Department, UK, 2002.
[18] Mynatt, E.D., *Making Ubiquitous Computing Visible*, College of Computing Graphics, Visualization and Usability Center, Georgia Institute of Technology, Atlanta, GA, 2000.
[19] Shneiderman, B. *Designing the User Interface: Strategies for Effective Human-Computer Interaction*, 5th edn, with C. Plaisant. Addison-Wesley Longman Publishing Co., Boston, MA, ISBN 0-321-26978-0.
[20] Hiroshi, I., *Bottles as a Minimal Interface to Access Digital Information*, Tangible Media Group, MIT Media Laboratory, Cambridge, MA, 2000.

Chapter 5

Wireless technologies for pervasive systems

Abstract

Ubiquitous computing and pervasive systems are strongly based on the communication between people and surrounding environment, better if in mobility. Wireless technologies make the needed mobile communication possible by freeing people and devices to be wired. Wireless communications had a fast development during past years, thus leading to the current low-cost devices that are more and more embedded in everyday items, such as cellular phones and other personal mobile devices. Owing to their key role in pervasive systems, this chapter presents some of the most used wireless technologies, giving some useful detail about them.

1 Wireless data transmission

Wireless technologies are rapidly evolving, and the enthusiasm of designers and users in using them is growing up consequently. Nevertheless, there are several problems due to the nature of the physical media used in wireless communications. In fact, it is hard to set up a comprehensive model of the media taking into account all possible interferences that lower the signal quality. The main goal of ongoing studies and researches on wireless technologies is a better knowledge of the radio spectrum. This will allow for the setup of rules for channel access and connection establishment, with the lowest interference level and with the minimum power requirements.

Wireless communications for data exchange take place mainly within the industrial scientific medical (ISM) band which is centred on 2.45 GHz, where

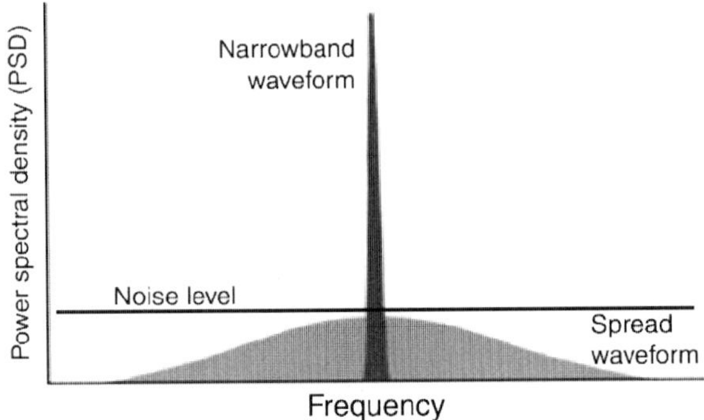

Figure 1: Power spectral density versus frequency (DSSS).

spread spectrum techniques are largely used to give them robustness against interferences.

The spectrum spreading on the whole ISM band is carried out in two main ways:

- The *direct sequence spread spectrum* (DSSS) is used in the Universal Mobile Telecommunications System (UMTS) for cellular phones. It reduces the power spectral density by modulating the signal with a well-known binary sequence at a higher rate (chip-sequence), thus making the signal itself more similar to a background noise (Figure 1).
- The *frequency hopping spread spectrum* (FHSS) transmits the signal in different time slots using a different carrier in each time, following a given hopping sequence (Figure 2).

It is hard to say what is the better way for digital data transmission over the ISM band. For sure, a DSSS modem is more complex than an FHSS one, but DSSS systems present a smaller overhead.

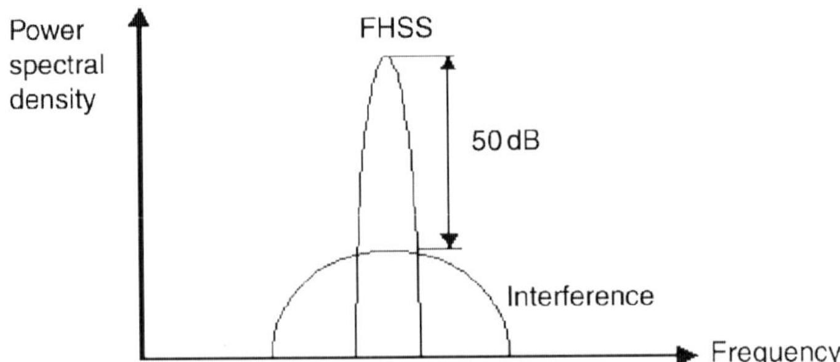

Figure 2: Power spectral density versus frequency (FHSS).

2 Bluetooth

Bluetooth is an industrial specification for exchanging data over short distances creating wireless personal area networks (WPANs). It represents a standard, secure and cost-effective method for information exchange among devices over a short-range radio frequency (RF) physical media.

Bluetooth was initially developed in 1994 by Ericsson [1] to allow mobile devices to communicate within a 30 feet (~10 m) range as a wireless alternative to RS232 data cables. In 1998, several majors in the technological field, such as Ericsson, IBM, Intel, Nokia and Toshiba, established the Bluetooth Special Interest Group (BT-SIG). From then on, more than 11,000 companies joined the BT-SIG and also all cellular phones and personal digital assistant (PDA) manufacturers.

Despite its unique features, Bluetooth is based on other previously existing wireless solutions, such as Motorola PIANO, IrDA, IEEE 802.11 and digital enhanced cordless telecommunications (DECT).

PIANO was designed to set up ad-hoc PANs. This feature was considered by the BT-SIG to improve the initial goal that envisaged Bluetooth only as a cable replacement. The full-duplex audio stream transmission is inherited from DECT and it is mainly used for wireless headsets. Raw data exchange is inherited from IrDA, whereas FHSS modulation within the ISM band, authentication, privacy and power management techniques come from IEEE 802.11.

The name is inspired by King Harald Blåtand of Denmark, an able diplomatic that joined all Scandinavian people together and introduced the Christian faith in the region. The Bluetooth inventors thought that this could be a suitable name for a technology aiming at the definition of a unique communication protocol among different devices (the same way as the king Blåtand joined people together).

Bluetooth is therefore a short-range wireless technology, now defined by the IEEE 802.15 standard, designed for data exchange among different electronic devices. Bluetooth has low energy requirements, thus making possible its integration within a large variety of devices, ranging from PCs and laptops to small peripherals, such as PDAs and cellular phones. Such devices can therefore interact with each another, thus allowing for audio/video exchange, internet connection sharing and all communication-based activities.

A Bluetooth radio can be embedded in a device or can be added afterwards, as for instance the pen drive-size radios with the USB hardware interface.

Figure 3: Master–slave setup in a piconet.

2.1 Piconets

To communicate with each another, Bluetooth units must be a part of small networks, called *piconets*, in which one unit plays the *master* role and the others act as *slaves* (Figure 3) [2].

The difference between master and slave is only at a logical level and it is tightly coupled with the piconet existence: when the piconet is loosened, master and slave roles are cancelled consequently. This means that there is no difference among Bluetooth devices at the physical level: all of them can act as both slave and master.

By definition, the master is the device that starts a piconet with other devices, which in turn become all slaves. Figure 3(a) shows a point-to-point connection between a master and exactly one slave, whereas Figure 3(b) shows a point-to-multipoint connection between a master and more than one slave. When a Bluetooth device establish a piconet (thus becoming a master), it sends its own parameters to other devices allowing them for the synchronisation.

All communications in a piconet can take place only between the master and one slave, no direct slave-to-slave communication is possible. The media access is managed by the master in a centralized way, by means of a polling scheme, in order to avoid collisions. A slave is authorized (and forced) to transmit in a given time slot only when enabled by the master with a suitable data packet sent during the preceding slot. Communications in a piconet take always place in both directions: when the master sends a packet to a slave, there is always an answer packet sent by the slave to the master.

A slave can show four operation modes:

- *Active mode.* The slave has a unique 3-bit address (the AMA, active member address). The '000' address is always assigned to the master and it is used for message broadcast to active members within the piconet. The number of active slave members of a piconet is therefore limited to seven. This restriction is not so strong if we consider the purposes and features of a small personal network. In its active mode, the unit actively participate in data exchange. During the master-to-slave slots, the unit listens to the channel: if packets contain its address, it keeps on listening until all packets are received and answers to the master in subsequent slots, otherwise it stops the listening until the next master transmission.

- *Sniff mode.* The slave listens to the channel with a lower rate, and it keeps on listening only if packets contain its address, otherwise it switches to sleep mode. The sniff mode is one of the three energy saving modes, the one that allow for the smaller energy saving.
- *Hold mode.* A slave can request to be in this mode or can be forced by the master. There is no data exchange while in hold mode, even if the device keeps its AMA. The only active process is the internal clock, thus allowing for a good energy saving.
- *Park mode.* The parked units have an 8-bit address (the PMA, parked member address), so there can be up to 256 units in this mode in a piconet. These units hold the master address and the clock, even if they do not participate in the piconet activities. Parked units periodically listen to a special channel (the *beacon* channel) for possible wake-up messages from the master. In this mode, the unit duty cycle is at the minimum, thus allowing for the higher energy saving.

Bluetooth devices can exchange both raw data and audio data, so there are two types of links can be established between master and slave: the *synchronous connection-oriented* (SCO) and the *asynchronous connection-less* (ACL).

The SCO link is a symmetric, point-to-point link between the master and a specific slave. The SCO link reserves slots and can therefore be considered as a circuit-switched connection between the master and the slave. The SCO link typically supports time-bounded information like voice. The master can support up to three SCO links to the same slave or to different slaves. A slave can support up to three SCO links from the same master or two SCO links if the links originate from different masters. SCO packets are never retransmitted. In this case a hand-shaking phase is needed, in which master and slave agree on synchronisation and on audio packet format to be used.

In the slots not reserved for SCO links, the master can exchange packets with any slave on a per-slot basis. The ACL link provides a packet-switched connection between the master and all active slaves participating in the piconet. Both asynchronous and isochronous services are supported. Between a master and a slave, only a single ACL link can exist. For most ACL packets, packet retransmission is applied to assure data integrity.

A slave is permitted to return an ACL packet in the slave-to-master slot if and only if it has been addressed in the preceding master-to-slave slot. If the slave fails to decode the slave address in the packet header, it is not allowed to transmit. ACL packets not addressed to a specific slave are considered as broadcast packets and are read by every slave. If there is no data to be sent on the ACL link and no polling is required, no transmission shall take place.

To better exploit the bandwidth, ACL packets can last one, three or five slots. In this way asymmetric connections can be made, using long lasting packets in one direction and short packets in the other.

Bluetooth uses a simple automatic retransmission system (the ARQ, automatic repeat request) to achieve the reliability for ACL links. According to this schema, after each packet sent in one direction, there must be an answer packet in the opposite direction. This contains a bit indicating whether the previous packet was correctly received (ACK, acknowledge) or not (NACK, negative acknowledge). In case of NACK answer or no answer at all, the sender transmits the packet again.

2.2 Establishing a Bluetooth connection

One of the main features of Bluetooth is the possibility to automatically search, find and connect other devices with no human involvement. Before starting to communicate, two devices must agree on the connection details, setting up the clock and the frequency hopping sequence, for instance. This agreement is achieved by means of *inquiry*, *scan* and *page* procedures [2].

Devices that are available for connections, periodically run the *scan* procedure, listening to possible *inquiry* or *page* messages coming from other devices. Devices searching for others to connect use the *inquiry* and *page* procedures. The first one is used to discover existing (and available) devices within the Bluetooth radio coverage area. The *inquiry* procedure also allows a device to know details about the connecting one, such as the physical address, the clock and the name (if it exists).

These information are then used during the *page* procedure to complete the connection. In the *page* phase, the calling device sends its details to the called one so that the connection can be correctly established. The calling device becomes the master and the called device becomes the slave, but both can agree to switch their roles at any time during the connection.

To improve the security of a connection, there can be a preliminary phase to authenticate devices one another, that is, the *pairing* phase. Pairs of devices may establish a relationship by creating a shared secret key (*link key*). If a link key is stored by both devices they are said to be *bonded.* A device that wants to communicate only with a bonded device can cryptographically authenticate the identity of the other device, and so be sure that it is the same device it previously paired with. Once a link key has been generated, an authenticated ACL link between the devices may be encrypted so that the data that they exchange over the airwaves is protected against eavesdropping. Link keys can be deleted at any time by either devices: if done by both devices, this will implicitly remove the bonding between the devices. It is therefore possible one of the device to have a link key stored but not be aware that it is no longer bonded to the device associated with the given link key.

2.3 Scatternets

As described earlier, there can be up to seven slaves at a time in a piconet communicating with one master. It is possible to increase the number of communicating devices by means of *inter-piconet units* (IPUs) that set up *scatternets*.

An IPU is a Bluetooth unit that is part of more than one piconet. Each IPU can communicate with one piconet at a time, since it has only one Bluetooth radio, but it can switch from one piconet to another from time to time, keeping clock and address of respective masters. Each IPU can act independently in the piconets (Figure 4(a)) or act as a gateway between the piconets, forwarding packets from one to the other (Figure 4(b)). A group of piconets in which connections consists between different piconets by means of IPUs is called a *scatternet*.

2.4 The Bluetooth stack

Besides all radio specifications, Bluetooth defines a layer protocol architecture consisting of core protocols, cable replacement protocols, telephony control protocols and adopted protocols.

Figure 5 shows the Bluetooth-layered stack, in which every layer within a device logically interacts with the peer layer in another device with the appropriate protocol. Physically, each layer supply services to the upper one exploiting services from the lower one, similar to the Open Systems Interconnection (ISO-OSI) model [3].

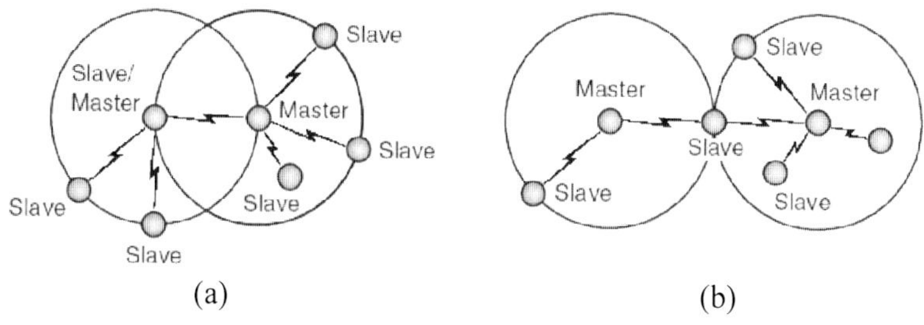

Figure 4: IPUs and Scatternets.

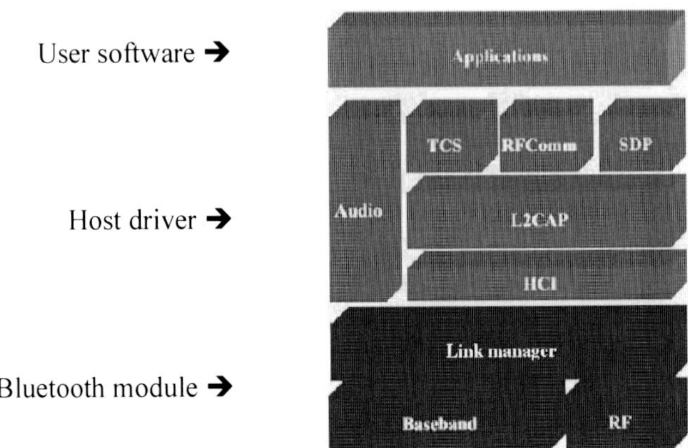

Figure 5: The Bluetooth protocol stack.

The *RF* layer is concerned with the physical data transfer, carrier generation and modulation, and power management. It includes design specifications to be followed for the construction of a Bluetooth transmitter, such as spurious emissions inside and outside the band, the accuracy in frequency and the interference among channels.

The *Baseband* layer manages physical connections and channels, offering a variety of services: Bluetooth units synchronization, selection of frequency hopping sequence, error correction, data flow control and security.

The main task of the *Link Manager* layer is to create, manage and terminate connections, to manage the master–slave role switch and low-power modes (hold, sniff and park). Furthermore, it is responsible for the quality of service, authentication and encryption.

The *L2CAP* layer (Logical Link Control & Adaptation Protocol) provides higher layers with services for sending data [4]. To this end it supports multiplexing for higher levels of protocol, segmentation and reassembly of large packets, and management of logical connections with the upper layer.

The *HCI* host–controller interface (HCI) defines standardised communications between the host stack (e.g. a PC or mobile phone OS) and the controller (the Bluetooth module) [5]. This standard allows the host stack or Bluetooth module to be swapped with minimal adaptation. There are several HCI transport layer standards, each using a different hardware interface to transfer the same command, event and data packets. The most commonly used are USB (in PCs) and UART (in mobile phones and PDAs). Figure 6 shows the lower layers of a Bluetooth device. Upper layers interact with the baseband using the commands made available by the HCI through the HCI driver.

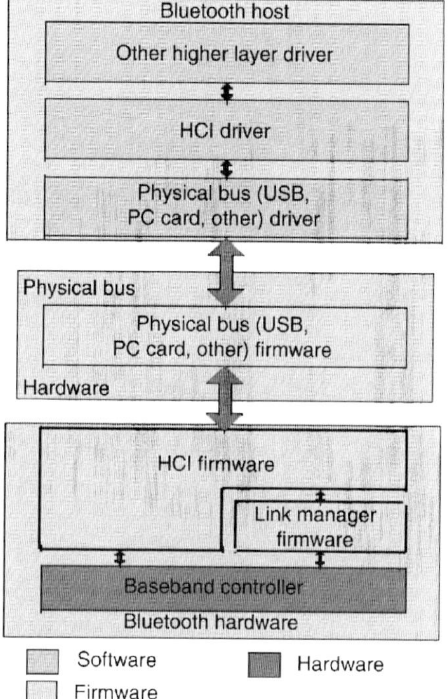

Figure 6: Data flow between a host and its Bluetooth device.

RFComm is a transport protocol that provides emulation of serial ports over the L2CAP layer.

The Service Discovery Protocol (SDP) layer provides a method for applications to discover which services are available and what are their characteristics. There can be different kinds of service, such as fax, printing and access to wired networks. However, SDP only provides a way to discover what services are available in a certain area, but does not allow access to them. To use the services offered by a particular server, there must be opened a connection to it. A network of Bluetooth devices is highly dynamic (some servers may leave the coverage range and others may enter at any time). For this reason, there is a notification mechanism that indicates when a new server is available, allowing the client to request the list of services it offers. When a server leaves the client range there is no notification, but the client can update the server list via SDP, thus cancelling those which may no longer be available. The service discovery can take place in two ways:

- The client is looking for a service with given attributes and wants to know if the server is able to provide a service that meets those requirements.

Since each service is identified by a universally unique identifier (UUID), in this case the search is carried out by providing the UUID. The UUID are universally recognized, hence they are not related to a particular server.

- The client is not looking for a particular service, but only wants to know what are those provided by the server. This discovery mode is named 'service browsing'.

2.5 Bluetooth profiles

Besides, the protocols that allow two units to communicate in a consistent manner, Bluetooth also defines some *profiles* that are associated with given applications [6]. The profiles identify which elements of the protocol are needed in different applications. In this way, devices with limited memory or computing resources, such as headsets or mouse, may only implement the protocols of interest for the applications they are intended. New profiles can be added to the Bluetooth specifications, following the introduction of new applications.

There are some well-known profiles corresponding to Bluetooth 'official services', and, in particular, the following profiles:

The *Generic Access Profile* (GAP) is the basis on which applications are built. Its main task is to provide a way to establish and maintain secure connections between the master and the slave. It sets appropriate specifications for the security level, for the user interface of all Bluetooth devices and for all the operational modes.

The *Service Discovery Application Profile* (SDAP) is used by devices to find out what services are offered by other devices. This profile can work either as a server (i.e. it can be queried by another device answering with their own services) and client (querying the other devices). Each unit has all the information about available services and supported protocols. Bluetooth devices may use this information to verify whether the interaction with other devices within the piconet is possible or not.

The *Serial Port Profile* (SPP) is a transport protocol that is used by most of the other profiles. It emulates a serial link and it is useful especially with legacy applications that need that kind of link.

The *Generic Object Exchange Profile* (GOEP) defines a client–server relationship for data exchange. Clients start the transactions, and a slave node can act as both client and server. As the SPP, this profile is a building block for other profiles.

The *LAN Access Profile* (LAP) allows a Bluetooth device to connect to a LAN.

The *Dial-up Networking Profile* (DUNP) allows a PC or laptop to connect a modem without any wire (e.g. a mobile phone equipped with a modem).

The *FAX Profile* is similar to the DUNP profile, enabling wireless fax machines to send and receive faxes using mobile phones.

The *Intercom Profile* (IP) allows two phones to connect as transceivers.

The *Headset Profile* (HSP) allows a headset to connect its base station.

The *Object Push Profile* (OPP), the *File Transfer Profile* (FTP), and the *Synchronization Profile* (SP) are devoted to the objects exchange between two wireless devices. Objects could be electronic business cards, pictures or data files. The SP profile, in particular, is designed to synchronize data between two devices, such as the phonebook of a PDA and a PC.

3 Wi-Fi

Wi-Fi (wireless fidelity) is currently the most used technology to implement wireless local area networks (WLANs). It is defined by the IEEE 802.11x standard [7].

802.11 WLANs are based on a cellular architecture in which the area where the service should be available is divided into cells, as well as in the GSM telephony networks. Each cell (called basic service set, BSS) is controlled by a base station also known as access point (AP) [8]. An AP is similar to a network hub, relaying data between connected wireless devices in addition to a (usually) single connected wired device, most often an ethernet hub or switch, allowing wireless devices to communicate with other wired devices. This is the *infrastructure* operational mode (Figure 7). Even if a simple wireless LAN can be composed of a single cell with a single access point, in most cases, there will be a number of cells in which access points are interconnected through some type of distribution network (which is usually defined *distribution system* or DS). A set of several interconnected WLANs, including various cells, the corresponding access points and the

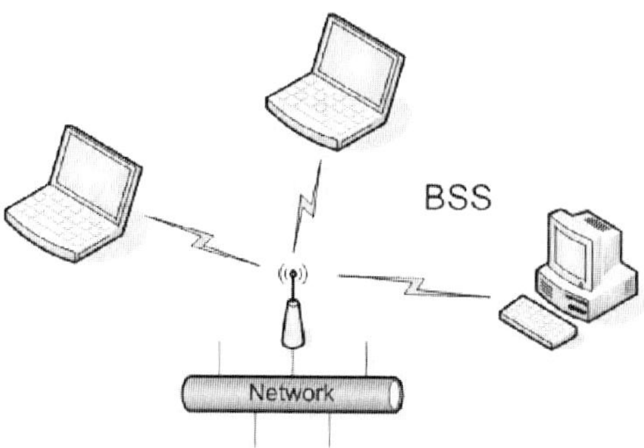

Figure 7: Wi-Fi WLAN – Infrastructure mode, single cell (BSS).

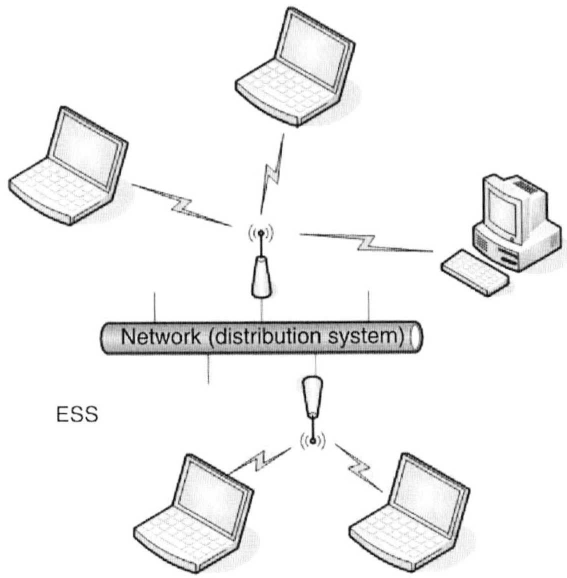

Figure 8: Wi-Fi WLAN – Infrastructure mode, multiple cells (ESS).

distribution system, is known as extended service set (ESS) (Figure 8). An ESS is seen as a single 802 network by higher levels of OSI model.

There is a different operational mode for Wi-Fi-based WLANs in which the access point is not needed. This mode is called ad-hoc, which allows direct connection to computers. In ad-hoc mode, wireless client machines connect to one another in order to form a peer-to-peer network, i.e. a network in which each machine acts as both a client and an access point at the same time. The setup formed by the stations is called the independent basic service set (IBSS) (Figure 9).

To access an existing BSS, each station needs to acquire synchronization information from the corresponding AP. This information can be acquired in any of the following ways:

- *Passive scanning*. In this case, the station waits for a *Beacon Frame* from the AP. A beacon frame is periodically transmitted by the AP, and it contains information about the timing of data transmission.
- *Active scanning*. In this case, the station looks for an AP by transmitting a *Probe Request Frame*, and then it waits for a *Probe Response Frame* coming from an AP.

Both methods are suitable and the choice between one or the other is carried out according to the energy or performance constraints.

The process that allows a station to move from a cell (or BSS) to another with no loss of connection is called *Roaming*. This function is similar to that

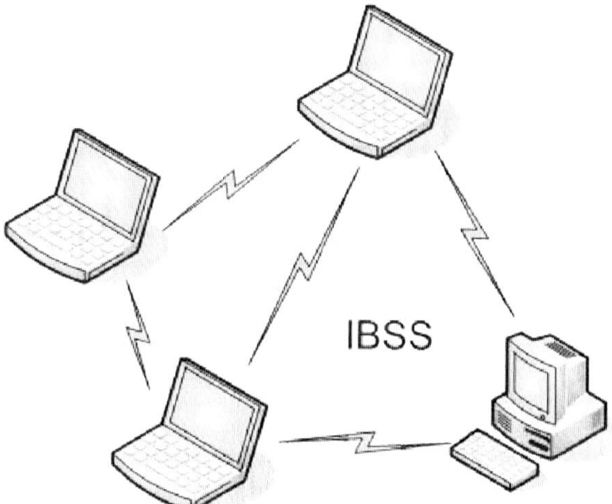

Figure 9: Wi-Fi WLAN – ad-hoc mode (IBSS).

which is done in cellular telephony systems. The 802.11 standard does not define how roaming should be done, but defines a basic operational mode.

The moving station detects which APs are available for connection carrying out the active or passive scanning. Depending on the received signal strength, it decides which AP is more convenient to join. Then the station uses a mechanism of re-association defined by the standard, by which it can cancel the association with the old AP and join the new one. The reassociation process consists of an information exchange between the two AP involved in the roaming through the distribution system, with no overload for the radio channel.

The possibility to roam from a cell to another along with a suitable ESS setup allows providers to supply people with wireless pervasive services in hotels, train stations, airports, government offices and shopping centres.

A further evolution of the Wi-Fi wireless technology is the IEEE 802.16 standard (Wi-MAX), which features are closer to a wide area network (WAN). The Wi-MAX consortium has several members, such as Intel, Siemens, Alcatel, Fujitsu, Sumitomo Electric, as well as telecommunication companies such as British Telecom, France Telecom and Qwest [9]. These ones are particularly interested in Wi-MAX, since it enables Internet service providers to allow for a broadband Internet access at a lower cost.

3.1 Technical details

IEEE 802.11 divides the available band into channels, analogously to how radio and TV broadcast bands are carved up but with greater channel width and overlap. Besides, specifying the centre frequency of each channel, 802.11 also specifies a spectral mask defining the permitted distribution of power across each channel. The mask requires that the signal can be attenuated by at least 30 dB from its peak energy at ±11 MHz from the centre frequency, the sense in which channels are effectively 22 MHz wide. One consequence is that stations can only use every fourth or fifth channel without overlap [10].

IEEE 802.11 – amendments b and g – uses the 2.4 GHz ISM band. Because of this choice of frequency band, 802.11b and g equipment may occasionally suffer interference from microwave ovens, cordless telephones and Bluetooth devices. 802.11b/g uses the DSSS signalling and orthogonal frequency-division multiplexing (OFDM) methods, respectively.

OFDM is a scheme utilized as a digital multi-carrier modulation method. A large number of closely spaced orthogonal sub-carriers are used to carry data. The data is divided into several parallel data streams or channels, one for each sub-carrier. Each sub-carrier is modulated with a conventional modulation scheme such as quadrature amplitude modulation (QAM) or phase shift keying (PSK) at a low symbol rate, maintaining total data rates similar to conventional single-carrier modulation schemes in the same bandwidth.

The primary advantage of OFDM over single-carrier schemes is its ability to cope with severe channel conditions – for example, narrowband interference and frequency-selective fading due to multipath – without complex equalization filters. Channel equalization is simplified, because OFDM may be viewed as using many slowly modulated narrowband signals rather than one rapidly modulated wideband signal. The low symbol rate makes use of a guard interval between symbols affordable, making it possible to handle time-spreading and eliminate inter-symbol interferences. This mechanism also facilitates the design of single-frequency networks, where several adjacent transmitters send the same signal simultaneously at the same frequency, as the signals from multiple distant transmitters may be combined constructively, rather than interfering as would typically occur in a traditional single-carrier system.

802.11a uses the 5 GHz band, which offers at least 19 non-overlapping channels rather than the 4–5 offered in the 2.4 GHz ISM frequency band, and data transfer rates up to 108 Mbps.

4 IrDA

Infrared Data Association (IrDA) [11][12] was designed to allow for simple short distance communications (<1 m). Devices involved in such communication need to be in sight, and this is one of the main IrDA disadvantages. Besides, the short range, the angle between the infrared emitter and the receiver must be less than 30°. Despite this important drawback, IrDA is the current standard for infrared connections, being very common in notebooks, PDAs, cellular phones and other mobile devices.

The data transfer rate of IrDA connections can be up to 4 Mbps, but it can reach 16 Mbps with the latest version of the standard (fast infrared, FIr).

IrDA can be also used for serial connections (serial infrared, SIr). In this case, it can reach a maximum data transfer speed of 155 Kbps.

The intrinsic drawbacks of IrDA, mainly due to its short operational range, are often exploited as security features. In fact, it is almost impossible to hack an IrDA connection with active or passive attacks (such as man-in-the-middle or sniffing), unless the hacker is very close (and within the sight constraints) to the attacked hosts.

5 HomeRF

HomeRF is a wireless technology designed for domestic applications. It is a standard developed by the HomeRF Working Group, which is also part of the Bluetooth SIG [13], and it uses the same frequency band as Bluetooth and Wi-Fi (ISM 2.4 GHz). HomeRF is supported by big companies such as Motorola, Intel and Compaq, all of them are producers of digital communication devices for home applications.

HomeRF is based on the Shared Wireless Access Protocol (SWAP), which allows for six voice channels (according to the DECT standard) and a Wi-Fi data channel. The first versions of these devices had a maximum data transfer rate of 2 Mbps, but the newer appliances achieve higher transfer rates, up to 10 Mbps, by means of the SWAP version 2.0.

HomeRF uses the FHSS modulation to increase security and reliability, and to reduce interferences. Since it does allow for ad-hoc point-to-point connections only, there is no need of APs and the needed hardware is quite simple. This makes HomeRF a cheap choice for home wireless networks. Lower costs and simpler hardware have a negative effect on the covered range (less than 30 m) and interoperability with any existing wired network.

6 Wireless technologies comparison

The comparison of above-discussed technologies is not easy, because some technologies are complementary rather than competing.

Table 1 shows a comparison of wireless technologies for home automation, indicating the applications they were originally designed for, the modulation, the maximum bit rate and the standard issuers.

Table 1: Wireless technologies main features.

Name	Application	Features	Modulation	Max rate	Issuer
802.11	WLAN	Broad band, high speed	DSSS (802.11b)	11 Mbps	IEEE
			OFDM (802.11a/g)	6–108 Mbps	
HomeRF	SOHO	Low cost	FHSS	1 Mbps (SWAP 1.0)	HRFWG
				2 Mbps (SWAP 2.0)	
Bluetooth	Cable replacement	Broad band, low cost	FHSS	1 Mbps	BT-SIG

Table 2: Wireless technologies comparison.

Name	Max rate	Rel. cost	Net supp.	Range	Currents
802.11	108 Mbps	Med/Hi	TCP/IP	> 50 m	10–400 mA
HomeRF	2 Mbps	Medium	TCP/IP	> 50 m	1–300 mA
Bluetooth	1 Mbps	Med/Low	PPP	< 10 m	1–60 mA
IrDA	16 Mpbs	Low	PPP	< 2 m	10 μA to 300 mA

Table 2 is the summary of described wireless technologies. It shows the maximum bit rate, the relative cost of devices, the supported protocols for integration with data and/or voice networks, the operational range and the approximate power consumption.

7 RFID

At the end of this chapter, we decided to spend a paragraph for radio frequency identification (RFID), even if it is a wireless technology not intended for communications. Owing to its features, RFID became one of the most used wireless technology in pervasive systems implementations. In fact, due to its flexibility, it can be used in several different ways than those it was designed for.

RFID technology is a contactless identification that can read labels which hold data. RFID was initially developed for items detection and access control applications. It is particularly effective in industrial production and, more generally, in environments where bar-coded labels are difficult to read. RFID can detect moving objects, so it is widely used in the transport management field, for the automatic identification of vehicles and cargo.

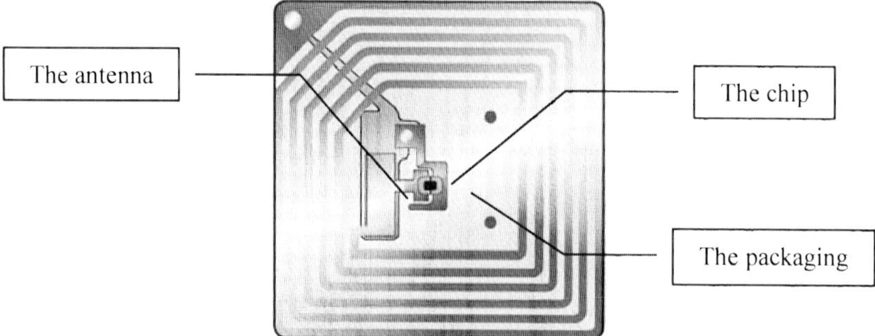

Figure 10: Components of a RFID tag.

A basic RFID system consists of two components: a reader and a transponder (tag) [14][15]. The tag may be electronically programmed with static data (read-only tags) or variable (read/write tags). The tags consist of a chip (1 mm^2, 0.2 mm thickness) connected to an antenna glued on a thin layer of polyethylene terephthalate (PET), which can be inserted into labels, tickets, bracelets, buttons, keys, etc. (Figure 10). All tagged objects become *smart* since held data can be used to build some additional information (e.g. by querying a database).

Tags can be classified by several criteria, such as the energy source, the memory type, the working frequency, and so on. We will shortly discuss the energy source that classifies tags in *active* and *passive*.

7.1 Passive tags

Passive tags are the most popular and cheap because they do not embed a battery or any other power source. Besides, receiving and transmitting data within a short range (25/120 cm), the antenna turns the received electromagnetic energy into electric energy to power up the transmitter. This is the reason why the range of a passive tag depends on the amount of received energy and therefore on the reader transmitted power.

The intelligent part of each tag consists of a simple RF signal transmission circuit and a non-volatile memory containing a unique code, which is transmitted to the reader. Some tags embed a more complex chip which is able to carry on some computation and send back the results to the reader.

Most of the passive tags operate in the RF range of 13.56 MHz or 125 KHz, and their size can be very small as well as their cost, which is much lower than active tags.

7.2 Active tags

Active tags are equipped with an internal power source, often a small lithium battery. This is used to supply power to the transceiver and to keep alive the content of a static RAM, where tag data are stored. The use of an internal power supply enables the tag to operate at higher frequencies and higher transmission power, thus reaching a longer operational range up to 15–20 m. These types of tags are used for automatic identification of moving objects, such as cars along the highway, where the detection distance can be quite long and not fixed. Most of the existing active tags operate within the 450 and 900 MHz bands, for which there is a wide range of cheap electronic components commonly available.

7.3 Readers/writers

Despite its name, an RFID tag reader is a device that can read (decode) and write (encode) data from/to a transponder. Besides these basic activities, a reader is often part of a LAN in order to receive data to be written on the tag or to forward data read from the tag.

A reader is normally composed of a control unit (the *controller*), a network controller with different interfaces (serial ports RS232/RS422/RS485, Ethernet, Fielbdus, Profibus, etc.), and one or more antennas to exchange data with the tags. An external dedicated antenna allow for better performances, but smaller systems, such as the Compact-Flash ones, have small integrated antennas.

The reader plays a key role within an RFID system. In fact, it must deal with all radio communications with all the tags that are within its range, correctly managing data exchange with many tags simultaneously, and avoiding or resolving collisions, if some occurs.

RFID tag readers may be fixed (production lines, moving objects control and detection), transportable (installation on trucks, forklifts, etc.), or hand-held to be integrated within a PDA for activities 'in the field'.

7.4 RFID systems

An RFID system consists of a transceiver (reader) and one or more transponders communicating one another via a frequency-modulated radio signal. The transponders receive the RF signal from the reader and then send their data back to the reader by modulating an RF signal. In the case of passive tags, the energy needed for their operation is produced by the same RF received signal. Data sent to the reader is often a unique code, chosen among several billion of possible combinations, which is stored on the chip during its production.

A system like that is very effective in a large variety of application fields, especially if it is used in combination with a suitable network-based data collection and processing system.

RFID systems are particularly useful when a contactless item detection is needed, along with a high hit-ratio (the correct readings on the first try is more than 99.5%), even in dirty or particularly severe environments. Last but not least, RFID systems can be easily hidden in the environment, so they are useful when the presence of a control system must not be revealed.

7.5 RFID for pervasive systems

RFID technology can significantly contribute to the realization of pervasive systems, as demonstrated by many researchers over the years. Some examples are the magic medicine cabinet [16], the augmentation of desktop items [17] and smart shelves [18]. These prototypes show that RFID technology has many benefits over other identification technologies, because it does not require line-of-sight alignment, multiple tags can be identified almost simultaneously, and the tags do not destroy the integrity or aesthetics of the original object. Owing to the low cost of passive tags and their powerless operation, there are also some weaknesses associated with RFID-based object identification, as shown in [19].

The use of RFID applications exploit the full range of tag technology, from low-cost tags to highly expensive miniature sensor/transponders. Animals and livestock have been tracked using RFID technology for decades. Recently, RFID has become a technology of choice for tracking humans too, by means of smart tags the size of credit cards, buttons, bracelets, and even tiny chips embedded in the skin [20][21].

Many researchers have demonstrated over the years that RFID technology can significantly contribute to the realization of augmented reality systems, where technology is seamlessly integrated into the environment.

Owing to its cost-effectiveness and easiness of use and deployment, RFID is more and more used as enabling technology to provide humans with useful services for their everyday lives, ranging from interactive guides [22][23], to indoor navigation systems [24], and social interaction [25]. Most existing solutions, moreover, employ a single specific technology, or are based on simple querying dedicated web pages.

Acknowledgements

This chapter was written with the contribution of the following students who attended the lessons of 'Grids and Pervasive Systems' at the faculty of

Engineering in the University of Palermo, Italy: Aulico, Caruso, Desiderato, Rossini and Scelfo.

References

[1] The Bluetooth Special Interest Group official website, *About the Bluetooth SIG*, https://www.bluetooth.org/About/bluetooth_sig.htm, accessed on May 2009.
[2] The Bluetooth Special Interest Group official website, *Communication topology*, http://www.bluetooth.com/Bluetooth/Technology/Works/Communications_Topology.htm, accessed on May 2009.
[3] The Bluetooth Special Interest Group official website, *Core system architecture*, http://www.bluetooth.com/Bluetooth/Technology/Works/Core_System_Architecture.htm, accessed on May 2009.
[4] The Bluetooth Special Interest Group official website, *L2CAP*, http://www.bluetooth.com/Bluetooth/Technology/Works/Architecture_Logical_Link_Control_and_Adaptation_Protocol_L2CAP.htm, accessed on May 2009.
[5] The Bluetooth Special Interest Group official website, *Host–Controller Interface*, http://www.bluetooth.com/Bluetooth/Technology/Works/Architecture__Host_Controller_Interface_HCI.htm, accessed on May 2009.
[6] The Bluetooth Special Interest Group official website, *Profiles Overview*, http://www.bluetooth.com/Bluetooth/Technology/Works/Profiles_Overview.htm, accessed on May 2009.
[7] IEEE 802.11 Wireless Local Area Networks, *The Working Group Setting the Standards for Wireless LANs*, http://www.ieee802.org/11/, accessed on May 2009.
[8] *IEEE Std 802.11-2007*, Section 3.16, p. 6, June 2007, http://standards.ieee.org/getieee802/download/802.11-2007.pdf, retrieved on May 2008.
[9] The WiMAX Forum, *WiMAX Forum Overview*, http://www.wimaxforum.org/about, accessed on May 2009.
[10] Cisco Systems, *Channel Deployment Issues for 2.4 GHz 802.11 WLANs*, http://www.cisco.com/en/US/docs/wireless/technology/channel/deployment/guide/Channel.html, retrieved on February 2007.
[11] IrDA Official Website, http://www.irda.org, accessed on May 2009.
[12] Knutson, C.D. & Brown, J.M., IrDA Principles and Protocols, 2004, ISBN 0-9753892-0-3.
[13] HomeRF Archives, http://www.cazitech.com/HomeRF_Archives.htm, accessed on May 2009.
[14] How RFID works, http://www.howstuffworks.com/rfid.htm, accessed on May 2009.
[15] Association for Automatic Identification and Data Capture Technologies, "What is RFID?", http://www.aimglobal.org/technologies/RFID/what_is_rfid.asp, accessed May 2009.

[16] Wan, D., Magic medicine cabinet: A situated portal for consumer healthcare. *Proc. Int. Symposium on Handheld and Ubiquitous Computing*, Lecture Notes in Computer Science, Vol. 1707, p. 352, 1999.
[17] Want, R., Fishkin, K.O., Gujar, A. & Harrison, B.L., Bridging physical and virtual worlds with electronic tags. *Proc. ACM CHI'99*, May 15–20, Pittsburgh, PA, pp. 370–377, 1999.
[18] Decker, C., Kubach, U. & Beigl, M., Revealing the retail black box by interaction sensing, *Proc. 23rd Distributed Computing Systems Workshops (ICDCS 2003)*, May 19–22, pp. 328–333, 2003.
[19] Floerkemeier, C. & Lampe, M., Issues with RFID usage in ubiquitous computing applications. *Pervasive Computing: Second International Conference, PERVASIVE 2004*, April 18–23, pp. 188–193, 2004.
[20] Sullivan, L., Legoland uses wireless and RFID for child security, *InformationWeek*, April 28, 2004.
[21] Purohit, C., *Technology Gets under Clubbers' Skin*, CNN, June 9, 2004, http://edition.cnn.com/2004/WORLD/europe/06/09/spain.club/, accessed May 2009.
[22] Po Yu Chen, Wen Tseun Chen & Cheng Han Wu, A group tour guide system with RFIDs and wireless sensor networks, *Proc. 6th Int. Conf. On Information Processing in Sensor Networks*, Cambridge, MA, pp. 561–562, 2007.
[23] Wang, Y., Yang, C., Liu, S., Wang, R. & Meng, X., A RFID & handheld device-based museum guide system. *2nd Int. Conf. on Pervasive Computing and Applications*, pp. 308–313, 2007.
[24] Miah, Md. Suruz & Gueaieb, W., An RFID-Based robot navigation system with a customized RFID tag architecture. *Internatonal Conference on Microelectronics*, 2007, pp. 25–30.
[25] Willis, K.S., Struppek, M., Chorianopoulos, K. & Roussos, G., *Shared Encounters, CHI 2007 Proceedings, Workshop Extended Abstracts*, pp. 2881–2884, 2007.

Chapter 6

Positioning in pervasive systems

Abstract

Positioning is a featuring topic of pervasive computing and context-aware services providing. Positioning plays a relevant role in coupling what to do and context factors such as who, where and when. Other factors typically deal with semantics and allow a system to arrange those services someone may expect in a given reality. Positioning can contribute to better instantiate a context by providing time and space information. Time information is often simple to get; on the contrary, the position of people who need to be provided with context-aware services is unpredictable and more difficult to detect because of the use of wireless communication. In this chapter we will give an overview of most used positioning methods, along with some existing implementations.

1 Introduction

Position-based services are becoming a relevant topic in context-aware applications. Location awareness is a basic requirement for new applications on mobile devices, as in the case of advertisement systems in large stores [1] and guidance systems in museums [2] with handheld devices which are only feasible with an accurate position estimation of the mobile terminal. There are many existing indoor and outdoor positioning systems which use different technologies and algorithms for positioning.

The first step to accomplish the positioning task is the distance estimation between a mobile device to be positioned (mobile device) and a number of devices whose location is known (reference devices). Distances can be estimated by means of several approaches [3].

Time-of-flight method estimates the distance between a moving object and a fixed point by measuring the time a signal takes to travel between the object and point *P* at a known speed. This method requires accurate clocks, especially for radio frequency (RF) signals. In this case, a 1-μs error in timing leads to a 300-m error in distance estimation.

The *attenuation* approach takes into account the fact that the intensity of an emitted RF signal decreases as the distance from the emission source increases. Given the signal strength at the emission, it is possible to estimate the distance from an object to some point *P* by measuring the signal strength at the point *P*. Of course, a function correlating attenuation and distance is needed.

Once distance has been estimated, the positioning task is mainly accomplished by the *triangulation* method. With this method, the position of an object is obtained by measuring its distance from multiple reference positions. 2D triangulation algorithm requires three different measurements from three non-collinear points, and the mobile device is located at the intersection of three circles (Figure 1). 3D positioning requires distance estimation from four non-coplanar points. For example, a global positioning system (GPS) receiver needs at least four satellites for 3D position estimation.

In the case of RF signal attenuation method, the circles' radius is evaluated on the measured strength of received signals, exploiting the relationship between signal strength and distance. RF signals are affected by a high degree of uncertainty (and consequently estimated distances too), and actually measured RF signal strength define a cloud-shaped region around the base station (Figure 2).

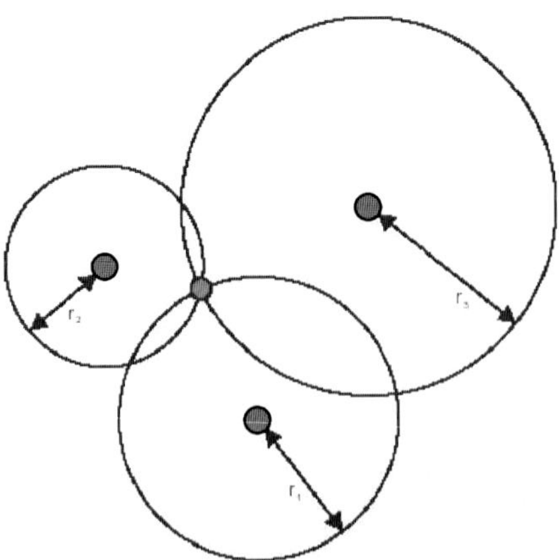

Figure 1: 2D positioning by triangulation.

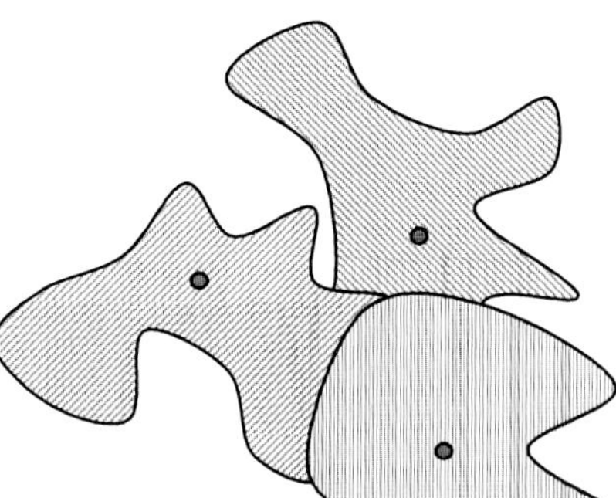

Figure 2: Actual RF signal coverage.

Regardless of the distance estimation approach used, most of the existing indoor and outdoor positioning systems need a number of reference nodes whose position is known to accomplish the positioning estimation task.

2 Position detection techniques

The three main techniques for automatic detection of the position are *triangulation*, *scene analysis* and *proximity*. Positioning systems can use them separately or in combination. In this section we will describe the basic concepts of each method and some existing deployment. Positioning systems cited in this section will be described in more detail in the next section.

2.1 Triangulation

The triangulation method uses the geometric properties of triangles to calculate the position of an object. It is divided into two subcategories: *lateration* and *angulation*. The first one uses distances measurements and the second uses angles measurements.

2.1.1 Lateration

The lateration technique calculates the position of an object by measuring the distance between the object and reference points. In a 2D space it is necessary to measure the distance between the object and three non-collinear points. In a 3D space it is necessary to measure the distance between the object and four non-coplanar points. The domain knowledge can reduce the number of required measurements. For example, the Active Bat positioning system [4] measures the distance between mobile tags (*Bats*) within a building and a grid of ultrasonic sensors placed on the ceiling. The 3D position of a Bat can be determined by only three distance measurements, since the sensors placed on the ceiling are always above the receiver.

There are three main approaches for distance measurement:

- *Direct measurement* is the measurement of distances using an action or a physical movement. For example, a robot may extend a probe until it touches an object to measure the distance from it. The direct measurement is simple to understand but difficult to achieve in an automatic way because of the complexity involved in the coordination of physical movements of an automaton.
- *Time-of-flight* method estimates the distance between a moving object and a fixed point by measuring the time a signal takes to travel between the object and point P at a known speed. In more detail, the mobile device sends a signal to a reference device, which in turn sends it back to the mobile device. Then the mobile device measures the round-trip time (RTT) of the signal. This leads to a circle, whose radius corresponds to half of RTT and whose centre is on the reference device. Therefore, a position estimation of the mobile device can be obtained by measuring three circles at least and by calculating their intersection.

  For example, an ultrasound impulse emitted from an object, travelling at 344 m/s, takes 29 s for the RTT to a fixed reference point P. This means it takes 14.5 s to arrive to the point P allowing us to conclude that the object is 5 m away point P.

  The use of electromagnetic waves such as RF ones in time-of-flight measurements is possible but requires very accurate clocks. In fact a light or RF pulse travelling at a speed of 299,800 m/s, will run 5 m in 16.7 ns.

  Another issue related to time-of-flight is the synchronization. When a single measurement is needed, as in the RTT of sound waves or radar reflections, synchronization is easy to achieve since the transmitter is the same as the receiver. In the GPS [5] the receiver is not synchronized with the satellite transmitters, so it cannot exactly measure the time that signal takes to reach the Earth from space. Therefore, the GPS satellites are synchronized with each other and send their local clocks along with the pulses thus allowing receivers to estimate the time-of-flight. As a consequence, GPS receivers can calculate their position in 3D space (latitude, longitude and altitude) using four satellites. The satellites are always above the receivers so only three satellites could be used to provide distance measurements to estimate the position in a 3D space. The GPS receiver requires a fourth satellite measurement to know the time shift between its local clock and the satellite clocks.
- *Attenuation* method takes into account that the intensity of a signal decreases as the distance from the source increases. Given a function

which links the attenuation and the distance for a type of signal, if the intensity (e.g. the RF transmitted power) of a signal is known, it is possible to estimate the distance between an object and any point *P* by measuring the strength of the signal received in *P*.

In environments with several obstacles, the measurement of distances using the attenuation method is usually less accurate than the one achieved with time-of-flight.

The SpotON positioning system is based on attenuation measurements by means of cheap identifiers [6]. SpotON uses clusters of identifiers and the correlation between multiple measurements to reduce the propagation-related negative effects on measurements accuracy.

2.1.2 Angulation

The angulation technique is similar to the lateration one, but it uses angle measurements instead of distances to estimate the position of an object. In a 2D space angulation requires the measurement of three angles or two angles and one distance which can be the distance between two points taken as reference (Figure 3). In a 3D space, four non-coplanar angles or three non-coplanar angles and one length measurements are required.

The angulation method is often implemented by means of phase-locked high-directional antennas since it is based on the direction of the received signal. A reference device measures the signal angle of arrival, which is sent by a mobile device. Location can be estimated by triangulation if two reference devices at least perform measurements.

The VHF omnidirectional ranging (VOR) air navigation system is an example of implementation angulation-based positioning. VOR stations are VHF transmitters placed in known locations which continuously transmit two pulses simultaneously. The first signal is an omnidirectional reference that contains the station identifier. The second signal quickly covers 360° (such as a rotating beacon) at a speed such that signals are in phase with the magnetic north and out of phase by 180° with the south. By measuring the phase shift,

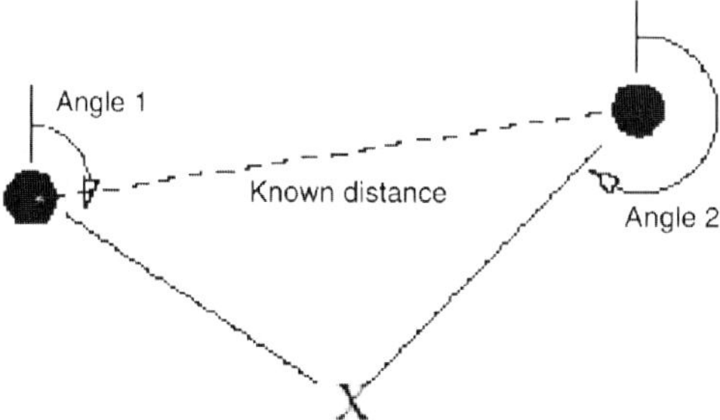

Figure 3: 2D angulation (two angles and a known distance).

the VOR receiver on a plane can compute the radial of the VOR station on which it is, i.e. the angle between the received signal and the magnetic north. The position of the plane can be calculated by using two VOR stations.

2.2 Scene analysis

The scene analysis is a technique for position detection that exploits the unique characteristics of a scene observed from a particular and appropriate point of view to identify the position of observer. Usually observed scenes are simplified to extract suitable features to be easily used and compared [7].

In *static* scene analysis, extracted features are compared with those archived in a dataset thus linking them to the observer's position. The *differential* scene analysis takes into account the differences between consecutive scenes to detect the position. The advantage of the scene analysis positioning method is that the position of an object can be inferred using passive observations and detecting features that are not angles or distances. The drawback of this technique is that the observer needs to know the environment features to be compared with those extracted from the observed scene. Furthermore, changes in the environment may alter the archived reference features thus requiring the construction of a new dataset.

The scene analysis is largely used in Mars exploration missions. Rovers first captured the surrounding environment by means of a photo camera mounted on the vehicle, thus constructing a reference scene map. During their exploration, rovers can detect their own position by comparing the current detected scene with those captured at the origin of the path [8].

Another existing positioning system based on the scene analysis method is the RADAR by Microsoft Research [9]. In this case the scene is not taken by a camera, but it is based on a RF map. In more detail, RADAR uses a dataset of RF measurements built by a mobile device which records the strength of received signals coming from fixed IEEE 802.11 stations. The position of a mobile device is then calculated by comparing the current received signal strengths with the reference ones based on the maximum likelihood principle.

2.3 Proximity

The proximity technique for position detection exploits the closeness of an object at a known position. There are three general approaches for proximity detection:

- *Detection by physical contact.* This type of proximity detection is carried out by means of contact sensors such as pressure, touch and capacitive field detectors. The capacitive field sensing is used for the implementation of the Touch Mouse [10] and for Contact which is a system for exchanging data between objects in direct contact with the human skin [11].

- *Monitoring of access points in a cellular network.* This method takes into account the presence of a mobile device within the coverage of one or more AP in a cellular network. Examples of implementation of this technique are the Active Badge [12] and Xerox ParcTAB [13], which use infrared cells in an office environment, and the Carnegie Wireless Andrew system [14], which uses a wireless radio cellular network.
- *Observation of automatic identification systems.* This detection method is based on the proximity to automatic identification systems such as credit cards readers, telephone records, access records to computers and RFID tags [15]. If the scan, query or monitor device is at a known position, then the position of the mobile object can be inferred.

3 Properties and features of positioning systems

In this section, we will present some properties and features that are often used to characterize and evaluate the positioning systems' performance and usefulness.

3.1 Physical vs. symbolic position

A positioning system can provide people with two types of information about their position: physical and symbolic. The GPS system, for example, provides physical information (or geometric) by means of latitude, longitude and altitude of a given place. On the other hand, a symbolic positioning system provides an abstract information about the position of a mobile object: 'in the living room', 'in Kuala Lumpur', 'at the Central Station', etc.

A system that provides physical information about a place can be enriched to provide information about a symbolic position too. For example, a system equipped with GPS can access a database which contains symbolic information for each physical location, as for instance in the automotive navigation systems. For this reason single abstract models for both representations have been studied, such as the one in [16]. The result of the integration is a hybrid, semi-symbolic model that decouples application representation from the sensor representation of location information. The hybrid model can accept location data in both forms and all data in the model can be viewed from both perspectives (Figure 4).

The accuracy of a physical positioning system may have implications over a symbolic information provisioning system. In fact, if a physical positioning system estimates a position with an error of 20 m, the symbolic position within a building can be wrong if rooms are smaller than 20 m.

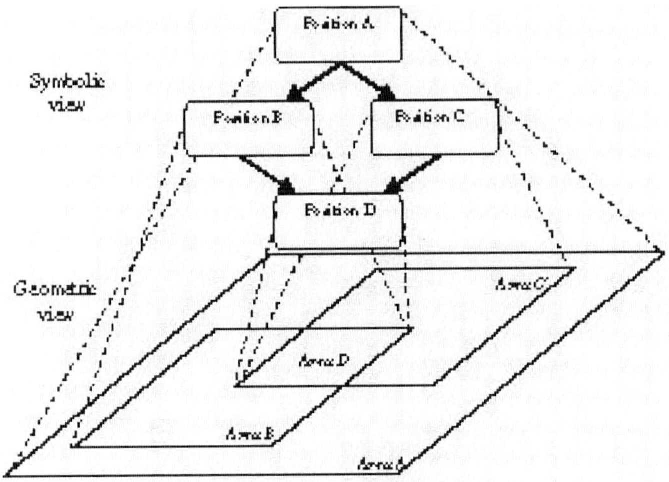

Figure 4: Semi-symbolic representation model.

3.2 Absolute versus relative position

Based on the reference points used by a positioning system, it can be classified as *absolute* or *relative*.

An absolute positioning system uses a universal grid reference to estimate the position of objects. For example, all GPS-based systems use the latitude, longitude and altitude to represent the position.

In a relative positioning system, each mobile object can have its own reference system. For example, during military actions, each soldier may know the position of companions as to his/her own position.

An absolute position can be transformed into a relative position (as to a second reference point) and vice versa. The triangulation is used for the determination of an absolute position from multiple relative readings if the absolute position of landmarks is well known.

However, often reference points position may not be known or reference points can be mobile themselves. As a consequence, when a positioning system is classified as absolute or relative, actually the classification is related to what information is available and how the system uses it, rather than any intrinsic property of the system.

3.3 Accuracy versus precision

A positioning system should return the position of objects to locate accurately and independently from the measurement units. Some inexpensive

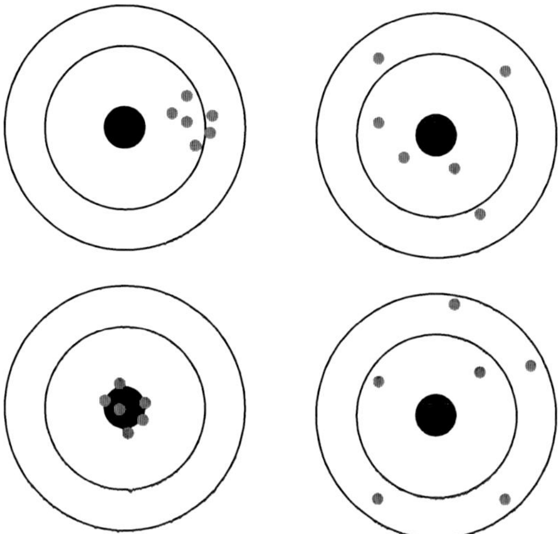

Figure 5: Difference between accuracy and precision.

GPS systems can locate an object within a radius of 10 m for approximately 95% of the measurements. Other more expensive units based on differential-GPS do much better, reaching an accuracy of 1–3 m in 99% of cases.

Distances characterize the *accuracy* of a system, whereas percentages indicate the *precision*, i.e. how often we can expect to achieve a certain level of accuracy. Figure 5 shows the difference between accuracy and precision: six shots have been fired on four targets. In A, the precision is good but the accuracy is poor. In B, the accuracy is quite good but the precision is poor. In C, both the accuracy and the precision are good. In D, both are poor. When comparing different positioning systems, both attributes have to be taken into account.

It is worth noticing that the accuracy itself is a relative attribute, so a system with a given accuracy can be not suitable for some application and suitable for another. For example, the accuracy required by a motion-capture system for computer animation is 1 cm or less, whereas the GPS-based systems used by biologists to locate flocks of migratory birds may have an accuracy of 1 km^2.

3.4 The range

A positioning system must be able to locate objects anywhere it is designed to operate: a single room, a particular building, a campus, a metropolitan area or the whole Earth. Moreover, the number of objects that a system can locate at the same time could be limited. For example, a GPS system can meet the demands of an unlimited number of receivers moving along the entire Earth using 24 satellites with the support of 3 more redundant satellites. Conversely, some readers of electronic tags cannot read any label if there is more than one in a certain radius.

As a consequence, to assess the range of a positioning system both coverage and number of objects it can locate per unit of infrastructure in a given time interval must be taken into account.

Positioning systems can increase their range by increasing their infrastructure. For example, a system used to locate objects in a single building can operate throughout a campus provided that all its buildings and its external areas are equipped with the required sensors and infrastructure.

3.5 Identification

Applications that need to identify or classify detected objects so as to establish a specific action based on their position require an automatic mechanism for identification. For example, a modern baggage sorting system at an airport should automatically distinguish luggage that must be directed to the belt for withdrawal or to the departing plane. This can be achieved through a system consisting of optical or RFID scanners placed at key points along the conveyor belts. The scanners read suitable labels on the baggage so that their identification is achieved. On the other hand, GPS satellites do not have appropriate mechanisms to identify the receivers.

Identification systems can recognize only certain types of features. For example, cameras and vision systems can easily distinguish the colour and shape of an object but cannot (easily) recognize individuals.

A general identification technique assigns names or globally unique IDs (GUIDs) to objects that the system have to identify. Every time a label on an object reveals its GUID, the infrastructure can access an external database to search for the object properties (name, gender, destination, etc.). It can also associate the GUID with other contextual information so that it can interpret the same object differently depending on the circumstances. For example, a person can find a description of an object in a museum in a particular language [17].

4 Positioning systems

After the short review about operating principles of positioning systems, we will now describe some of the most common existing positioning systems, evaluating them based on the features we discussed in the previous section.

4.1 GPS

The GPS was originally conceived by the American Ministry of Defence as a universal system to detect the exact position where a receiver is located on Earth and the exact time when the receiver is at that position.

The applications of GPS are not just limited to the military, but they are all also available for civilian use, albeit with some limitations in the achievable accuracy. The availability of GPS signals 24 hours a day in every corner of the globe and the progressive reduction in the costs of the receivers, turned the GPS into a social phenomenon, especially in the field of marine and air navigation.

The key features of the GPS are the following:

- The constellation was originally composed of 24 satellites and it was completed in 1993, of which 21 out of 24 were operational and the remaining 3 were for backup. Despite 24 is the minimum number of needed satellites due to orbital considerations, currently there are 31 satellites for redundant data transmission.
- The monitoring ground stations are placed in known positions. They transmit clock and ephemeris data to satellites, which, in turn, forward them to receivers along with navigation data.
- A GPS receiver calculates its position by precisely timing the signals sent by the GPS satellites high above the Earth. Each satellite continually transmits messages containing the time it was sent, precise orbital information (the ephemeris) and the general system health and rough orbits for all GPS satellites (the almanac). The receiver measures the transit time for each message and computes the distance to each satellite. Geometric trilateration is used to combine these distances with the location of the satellites to determine the receiver's location. The position is displayed, perhaps with a moving map display or latitude and longitude; elevation information may be included. Many GPS units also show derived information such as direction and speed, calculated from position changes.

The mean accuracy that is achieved with the common receivers is approximately 100 m in horizontal, 150 m in vertical and 340 ns in time. If the receiver is able to see more than four satellites simultaneously, it is possible to use mathematical algorithms to correct errors, thus increasing the accuracy. Errors are mainly due to shifts of satellites orbits, errors of satellites clocks and the effects of the ionosphere on the signal. Moreover, there is a random error added by GPS owner (US Department of Defence) for non-military users to avoid improper use of the system (such as targeting support for military or para-military attacks).

The GPS is not suitable for indoor positioning; the signal is too weak to penetrate inside a building. If a GPS-like system has to be used within large spaces, there can be used a local 'constellation' of transmitters (pseudolites, i.e. pseudo-satellites), to keep use the same receivers outdoors. As for the GPS, a minimum of four pseudolites must be received by the navigation system for suitable indoor GPS-based positioning.

4.2 Active Bat

The AT&T researchers developed the Active Bat positioning system using the lateration technique by means of ultrasonic time-of-flight [4].

In this system, users and objects to be located are equipped with small plates with radio-controlled ultrasonic pulse transmitters (the Bats). A local controller sends a short-range radio signal to the Bat. Once received the RF signal, the Bat sends back an ultrasonic pulse that is received by a grid of receivers mounted on the ceiling.

At the same moment in which the local controller sends the radio signal, the sensors on the ceiling start measuring the time interval between the radio signal and the received ultrasonic pulse, thus allowing the estimation of the Bat position. This task is carried out by a central controller by means of the trilateration technique. Of course, at last three sensors must be in range to carry on the position estimation task.

The nominal accuracy of Active Bat is 9 cm with a precision of 95%. The systems are suitable for identification purposes too, since each Bat has a unique identifier to be recognized.

The use of time-of-flight approach with ultrasonic fixed sensors requires a large infrastructure installed throughout the ceiling. Furthermore, the infrastructure must be placed in a suitable layout.

Discussions above show that Active Bat is a good positioning system concerning accuracy and precision, but it has short range, high infrastructural cost and difficulty in actual deployment.

4.3 RADAR

In early 2000s a research group at Microsoft has developed RADAR [9], a tracking system for large buildings based on the wireless technology W-LAN IEEE 802.11 which makes use of scene analysis technique.

RADAR is based on the principle that the received signal strength depends on the distance between the receiver and the transmitter. Actually RADAR uses both the received signal strength and the signal-to-noise (S/N) ratio to estimate the 2D position of a mobile device within a building.

It must be first created as a radio map of the covered area, that is, a database of sample places along with the received signal intensity from that places. A typical record of the radio map may be of the type (x, y, z, ss_i), where (x, y, z) are the coordinates of the place where the signal is recorded, and ss_i is the signal strength measured by the *i*th base station.

The RADAR approach presents two main advantages: it requires only a few base stations that have same infrastructure used to provide users with other kinds of wireless services.

Concerning the drawbacks, the first is that the object to be traced must be equipped with a W-LAN radio, and this could not be so common especially in smaller devices. The second disadvantage is that RADAR is not so expandable: the use of RADAR in multi-storey buildings or for 3D positioning is a hard task to be accomplished.

RADAR with scene analysis technique has an accuracy of 3 m with a precision of 50%. There is a version of RADAR that makes use of the lateration technique. In this case the accuracy is 4.5 m with the same precision of 50%.

Although the version that uses the analysis of the scene provides better accuracy, even little changes in the environment, such as the relocation of a cabinet, may cause the need to update the previous radio map or to build a new one.

4.4 MotionStar magnetic tracker

This system uses an electromagnetic-based method for tracing the position [18]. This type of technology is used by a wide range of products available for virtual reality and motion capture in the field of computer animation.

The MotionStar system [19] generates axial magnetic pulses through an antenna that transmits from a fixed position. The system calculates the position and orientation of receiving antennas by measuring the pulse response along three orthogonal axes, taking into account the ongoing effect of the Earth's magnetic field.

This tracking system allows for almost exact position and compass estimation. Its accuracy is less than 1 mm for the spatial resolution, 1 ms for the temporal resolution and 0.1° for compass resolution.

The main drawbacks of MotionStar are the excessive costs of implementation and the need to connect the object to be traced by a control unit. In addition, the range is quite short, since magnetic sensors must remain within 1–3 m from the transmitter. Lastly, the accuracy decreases with the presence of metallic objects within the environment.

Acknowledgements

This chapter was written with the contribution of the following students who attended the lessons of 'Grids and Pervasive Systems' at the faculty of Engineering in the University of Palermo, Italy: Collura and Tinnirello.

References

[1] Santangelo, A., Augello, A., Sorce, S., Pilato, G., Gentile, A., Genco, A. & Gaglio, S., A virtual shopper customer assistant in pervasive environments. *Lecture Notes in Computer Science: On the Move to Meaningful Internet Systems 2007*, OTM 2007 Workshops, eds. R. Meersman, Z. Tari & P. Herrero, Springer: Berlin, Vol. 4805/2007, 2007, ISSN: 0302-9743 (Print), 1611-3349 (Online), ISBN 978-3-540-76887-6, pp. 447–456, DOI: 10.1007/978-3-540-76888-3_67.

[2] Augello, A., Santangelo, A., Sorce, S., Pilato, G., Gentile, A., Genco, A. & Gaglio, S., A multimodal interaction guide for pervasive services access. *Proc. of 2nd Int. Workshop on Multimodal and Pervasive Services MAPS '07*, Istanbul, Turkey, 15–20 July, pp. 250–256, 2007, DOI: 10.1109/PERSER.2007.4283923.

[3] Hightower, J. & Borriello, G., Location sensing techniques, Technical report, *IEEE Computer Magazine*, pp. 57–66, August 2001.

[4] Harter, A., et al., The anatomy of a context aware application. *Proc. 5th Ann. Int. Conf. Mobile Computing and Networking*, ACM Press: New York, pp. 59–68, 1999.

[5] Getting, I.A., The global positioning system. *IEEE Spectrum*, **30(12)**, pp. 36–47, 1993.

[6] Hightower, J., Borriello, G. & Want, R., *Spoton: An indoor 3d location sensing technology based on RF signal strength*, http://seattle.intel-research.net/people/jhightower/pubs/hightower2000indoor/hightower2000indoor.pdf, accessed on May 2009.

[7] Ross Beveridge, J., Graves, C.R. & Lesher, C.E, Local search as a tool of horizon line matching. Image Understanding Workshop, Los Altos, CA, pp. 683–686, 1996.

[8] Orson, C.F., Probabilistic self-localization for mobile robots. *IEEE Transactions on Robotics and Automation*, **16(1)**, pp. 55–66, 2000.

[9] Bahl, P. & Padmanabhan, V.N. RADAR: An in-building RF-based user location and tracking system. *Proc. IEEE Conference on Computer Communications (INFOCOM)*, Tel-Aviv: Israel, Vol. 2, pp. 775–784, March 2000.

[10] Hincley, K. & Sinclair, M. Touch-sensing input devices. *Proc. 1999 Conf. of Human Factors in Computing Systems (CHI 1999)*, ACM, Pittsburgh, PA, pp. 223–230, 1999.

[11] Partridge, K., Arnstein, L., Borriello, G. & Whitted T. Fast intrabody signaling. *Demostration at Wireless and Mobile Computer Systems and Applications*, December 2000.
[12] Want, R., Hopper, A., Falcao, V. & Gibbons, J., The active badge location system. *ACM Transactions on Information Systems*, **10(1)**, pp 91–102, 1992.
[13] Schilit, B.N., Adams, N., Gold, R., Tso, M. & Want, R., The PARCTAB mobile computing system. *Proc. Fourth Workshop on Workstation Operating Systems (WWOS-IV)*, Napa, CA, IEEE Computer Society, pp. 34–39, October 1993.
[14] Bennington, B.J. & BartelWireless, C.R., Andrew: building a high speed, campus-wide wireless data network. *Mobile Networks and Applications*, **6(1)**, pp. 9–22, 2001, ACM/Kluwer special issue on wireless internet and intranet access, DOI: 10.1023/A:1009805518581.
[15] Want, R. & Russell, D.M., Ubiquitous electronic tagging. *IEEE Distributed Systems Online*, **1(2)**, 2000, DOI: 10.1.1.43.3680, http://www.parc.xerox.com/csl/members/want/papers/ubitags-con-2000R1.pdf.
[16] Magee, J., *Location Service for Mobile MultiMedia Environments*, Research Grant Final Report, http://www.doc.ic.ac.uk/~jnm/final-report.html, accessed on May 2009.
[17] Barton, J. & Kindberg, E T. *The CoolTown User Experience*, CHI2001 Workshop on Building the Ubiquitous Computing User Experience, http://www.hpl.hp.com/techreports/2001/HPL-2001-22.html, retrieved on May 2009.
[18] Raab, F.H., Blood E.B., Steiner, T.O., Jones H.R., Magnetic position and orientation tracking system. *IEEE Trans. Aerospace and Electronic Systems*, pp. 709–717, September 1979.
[19] Ascension Technology Co. website, *MotionStar*, http://www.ascension-tech.com/realtime/RTMotionSTARTethered.php, accessed on May 2009.

Chapter 7

Security in ubiquitous computing

1 Introduction

1.1 One single word: Security!

The goal of the ubiquitous computing (UC) is to provide mobile users an access to computing services through a wide area network like Internet. Even if most of the research has been addressed to the infrastructures and applications, security in UC is a difficult topic to manage.

The heterogeneous nature of the various devices, especially mobile devises, such as wireless laptops and personal digital assistant (PDAs), have given rise to various problems.

The aim is to obtain an effective access control to system resources and ensure the end-to-end security, without degrading the performance of the existing system. A security subsystem should be, in fact, quite efficient to run a computing session in real time, as soon as data arrive.

A security system should be resilient to a wide range of attacks [1]. To reach this goal it is necessary:

- to manage users and devices by appropriate authentication,
- to use an effective design of access system resources and
- to ensure a safe communication between two entities on the network.

Furthermore, we have to consider that in future we will have hundreds of computers connected to each other, cooperating through wireless networks. As a consequence, the management of security will become imperative.

1.2 Security in information systems

The use of computers in various fields of human activity has undoubtedly brought great benefits. In all areas where the main resources are information and the ability to process, computers are a useful and accurate storage and processing devices.

But as far as technology grows up, new problems emerge. A new instrument may be accepted only if the benefits that it offers are greater than the risks arising from its use.

In case of computers, there are a number of new security problems that should be adequately addressed and resolved.

A physical archive may be secured with bars on the windows and a security guard at the door that allows the access only to authorized personnel, but a database connected to the Internet introduces complex problems from a safety point of view. For example, in some systems it is easy to build programs that replace the login screen and steal passwords of other users; in a LAN is not difficult to obtain copies of the messages that transit on the network.

To solve these problems it is necessary to adopt specific security policies and implement the mechanisms applying them in some manner. It is important to separate policies from mechanisms. The policies state what needs to be done. Mechanisms determine how to realize them. A solution to the problem consists of an adequate security policy that, through the use of appropriate mechanisms, should ensure the protection of various resources from illegal accesses.

We can say that the objects to defend, essentially, belong to two conceptually separate categories but similar from an implementation point of view: physical resources and information.

While the UC will radically change the way we behave and interact between us and with IT systems, we must also take into account that this innovation has its risks.

UC will have an impact on society as similar, if not more, as the advent of the Web. But, like each innovation, it brings negative aspects also. It is therefore necessary to identify instruments and policies for protection before a critical mass of applications will be built and developed.

The traditional taxonomy identifies three main classes of security threats: confidentiality, integrity and availability [2].

Privacy is violated when there is an unauthorized access to protected information, like your medical records. Integrity is violated when there is an unauthorized change of information, such as when someone changes a sum or the beneficiary on a bank account. Availability is violated when the system is unable to perform the function for which it was intended, as when someone attacks a web site with the intent of turning it off.

These properties are all based on a distinction between authorized and unauthorized persons. The distinction between them involves a process that consists of three steps: identification (the user declares the identity), authentication (the system checks the validity of the request) and authorization (granting rights for the specific service). An authentication failure can easily lead to violations of confidentiality, integrity and availability.

For example, protecting our secrets with encryption is not very effective if the true identity of the recipient is different from what we expected. So it is natural to pay attention especially to the authentication process.

The safety requirements that a network must accomplish are as follows:

- *Authentication*: the users confirm their identity through questions that only they will be able to answer; two parties that want to communicate (exchange information) must first of all identify each other (mutual authentication);
- *Data integrity*: ensuring that the message has not been changed during the journey;
- *Secrecy*: encrypt data in a way that it is incomprehensible in case of their detection;
- *Access control*: access to resources must be controlled by and for the system and
- *Availability – confidentiality (privacy)*: a system must be available only to authorized users, only those who are authorized have access to confidential information.

The access to a computer always takes place through a channel of communication, whether virtual (remote login) or physical (an operator who sits in front of the console of a Mainframe has direct access to the machine). To realize the aforementioned threats, someone should access to the computer and then take control of one of the channels. In modern distributed systems connecting to a network through fast and flexible communication channels is a great way to access information, but at the same time it is an easy target for different types of attack, which essentially aim to examine/modify information which passes through them.

1.3 Transient secure association

Systems based on peer-to-peer today involve almost the entire world of computing, but their network connectivity is structurally not stable and therefore not guaranteed. Traditional approaches to authentication from Kerberos to Public-Key Certificates are therefore not enforceable, since they rely on online connectivity to an authentication server.

We need new solutions: these are of particular note in the transient secure association [2]. In the world of UC it would be desirable not disseminating remote controls for TV, stereo, DVD, VCR, curtains, central heating and air-conditioning through our houses. Instead, it would be interesting if all these systems obey to a one single universal remote control, which typically is generally a personal digital assistant (PDA) or something similar.

It won't be necessary to buy a remote with the device, you only need to establish an association between your PDA and the new device. Since you do not want your neighbour is able to activate your own devices, this association must be secure.

However this association could also be revoked (we must be able to sell our old stereo retaining our PDA and replace a broken PDA without losing control of all our devices), for this reason the Resurrecting Duckling [2] policy has been developed.

2 Security protocols

To secure wireless communications, protocols have been developed with the aim to guarantee a certain level of security for data transmitted and to give a control access to systems connected to a network.

A protocol is a set of rules that governs sequence and exchange of messages, or control signals, and the connection between devices.

There are different types of protocols that operate at different levels of the ISO/OSI stack, from the lowest to relate the simple management of electrical signals for communication to the highest such as security and authentication [3] that are going to analyse in greater depth.

2.1 Guarantees of a security protocol

A security protocol is a sequence of message exchanges between agents on an unsafe mean. Security protocols are usually executed before the communication protocols. However, this is not a rule.

According to the ISO standard, the main requirements that a secure network must satisfy [4] are: *confidentiality*, *integrity* and *non-repudiatebility*.

2.1.1 Confidentiality

Confidentiality includes

- *Confidentiality of data*: data sent and received should not be accessible to unauthorized users;
- *Confidentiality of the traffic flow*: prevents the acquisition of information from observation of the characteristics of data traffic and
- *Confidentiality of place*: ensures the confidentiality of the location of users.

2.1.2 Integrity

Information should not be editable by others. A message received by the recipient B should be identical to the original message sent by the sender A.

2.1.3 Non-repudiatebility

Non-repudiatebility includes

- *Non-repudiatebility of origin of a message*: agent B has a valid and irrefutable evidence that the agent A has been the one who sent the message and
- *Non-repudiatebility of receiving a message*: A has valid evidence that B has received the message.

2.2 Protocols developed for the security of wireless communications

Radio communications present an intrinsic problem of security. In a cable network generally data is not encrypted and once intercepted (sniffed) they are clearly visible. However, to enter into the network and to perform a sniffing of the traffic, you must physically connected to it and control the traffic flowing through the cables. In a wireless network instead, it is possible to listen to the communications that are taking place in the network through a proper radio equipment. 802.11 is a standard established by the IEEE, which provides the parameters of protection for this type of networks [5]:

- encryption with static wired equivalent privacy (WEP) keys and
- authentication WEP/EAP.

2.2.1 Encryption with static WEP keys

WEP is an encryption algorithm designed with the aim of making a wireless connection as secure as a connection via cable. According to this protocol on access point, two keys are preconfigured to 40 or 128 bits used by an algorithm, implemented at both ends, coding all the traffic in transit.

The main drawback of this system is the maintenance of the keys. If the key, that should be kept secret, is stolen, all the encrypted information is compromised. The more a key remains active, the more it is vulnerable. A peculiarity of the 802.11 standard is also the lack of a protocol for management of encryption keys, requiring that they should be manually handled on the various terminals. This limits the effectiveness of security systems, such as WEP.

2.2.2 WEP/EAP authentication

Extensible Authentication Protocol (EAP) is an extension created to make more secure the WEP protocol. EAP is based on dynamic change of keys. This makes the decrypting process more difficult to reach. Currently there are about 40 different ways to access the EAP. Access points and devices authenticate themselves at the beginning of each communication and generate the keys that will be used to encrypt the traffic only for the current session.

MAC address filtering is another way to protect network communications. The MAC address of a network adapter is a unique 12 digit hexadecimal code [GATES]. Since each card has its own unique address, it is possible to limit the access for the PA (access point) only to authorized devices with authorized MAC addresses, easily excluding anyone that should not be on the network. Several drawbacks prevent this system to offer a totally secure approach.

The first problem is the MAC addresses management. The wireless LAN administrator must keep up-to-date the database containing the list of the devices that have permission to access the network. This database must be kept on each AP individually or on a special radius server (a 'de facto' standard protocol for remote authentication) to which each AP is connected.

Each time a device is added, removed or modified in any way, the WLAN administrator must update the database of devices allowed. If this is limited to 10 or 20 people, it is not a problem, but in a corporate network with hundreds or thousands of devices this is certainly not a practical solution. Taking trace of the changes to the database would require a huge loss of time.

This heavy workload could be justified only if the MAC address filtering would be 100% secure. Unfortunately, the system is easy to defeat using the right tools. For example, using a wireless sniffer, an attacker can look at the traffic of the wireless network and can emulate MAC addresses of valid users, transmitted through the air, even if they are encrypted [GATES]. In this manner, security is compromised.

For small wireless networks, the MAC address filtering could be considered as a viable option in the absence of other security systems. For larger wireless networks, however, the simple MAC address filtering does not provide the level of security that could justify its enormous cost management.

2.2.3 Current status: the WPA, the best solution

With the growth of wireless networks, concerns have also increased about the security of data travelling via radio. What has been found is that wireless technology is not secure at all. As a matter of fact, packages and, therefore, information, travelling in a radius of several hundred meters. If they are not encrypted, anyone can see what we do.

At present, a protocol for secure connections is WEP, but there are hacking tools that can invalidate it. One example is AirSnort, a software that automates the cracking of the Protocol. AirSnort allows to 'sniffing' the wireless network traffic and get the master key to decode the encrypted data and gaining access to the wireless LAN (WLAN).

What to do? Wi-Fi Alliance has proposed a new standard security called Wi-Fi protected access (WPA) [DOORS] intended to replace the WEP in a short time. The new security protocol uses cryptographic algorithms and more robust authentication systems. It is capable to operate with current certified Wi-Fi products, and it is compatible with products already available on the market. And, most importantly, is able to resist to all types of attacks known so far. The WPA is based on a subset of features that are illustrated in specifications known as 802.11i Robust Security Network. WPA is able to run on all existing 802.11b hardware in the form of software update and will be able to ensure that data is protected and that only authorized people/ systems can access the network.

The everyday reality will be the test for this new security protocol. From a more technical point of view, the new WPA uses mechanisms of the future 802.11i standard for both data protection and for the process of users authentication. For data encryption, WPA uses Temporal Key Integrity Protocol (TKIP), which uses the same algorithm of WEP, but builds the keys in a different and more secure manner. With WPA each user has its own encryption key and this key can be changed periodically (dynamic keys and then no more static).

In corporate networks, the authentication process will be managed by a specific server capable to control WEP users in a simpler manner. For small house networks, with fewer demands in terms of safety, it is used a 'pre-shared key' modality, which has the advantage of not requiring the use of a server, even if the level of security is certainly lower.

The 802.11i standard will include two other types of cryptography: the Wireless Robust Authenticated Protocol (WRAP) and Counter with Cipher Block Chaining Message Authentication Code Protocol (CCMP) [GATES], which should bring the wireless network security to an optimal level.

As we know absolute security virtually does not exist, but it is certain that these passages of the Wi-Fi Alliance go in the right direction. Only when companies, organizations and home users will feel confident enough, wireless networks can definitely takeoff.

3 Encryption

The growing use of Internet as a medium for a rapid exchange of information, emphasized the need for secure, private and protected transmission of data from indiscreet glances.

Unfortunately, the network, as designed, does not support a good level of security and privacy. Information is transmitted in clear and it could be intercepted and read by anyone. Therefore, it has been necessary to create methods that make information indecipherable, ensuring therefore the integrity and enabling the authentication of the interlocutors, so that only the sender and the recipient can read and understand them.

Encryption algorithms try to protect, with a certain degree of security, high-value information against possible attacks by criminals, competitors or anyone who can use them to cause damage to someone else. This includes all aspects of safety messages, authentication of the parties and integrity verification.

Cryptographic systems have been present in history for centuries, especially for military reasons. Currently, encryption is no longer limited strictly to the military field. Research is currently trying to take all of the advantages of cryptography for creating a secure network available to the modern society.

3.1 Terminology

To better familiarize with the concepts in cryptography we give some definitions:

- *Encryption* is defined as the art or science of making secret messages.
- *Cryptographic analysis* is the art of violating a cryptographic system and decipher messages.
- *Cryptology* is the branch of mathematics that studies the mathematical foundations of cryptography.
- The clear message is called *plaintext* or *clear text.*
- The encrypted message is called *ciphertext.* It looks like a random sequence of symbols and it is incomprehensible.
- Encryption and decryption are the transformations of a message from plaintext to ciphertext back. They usually are generated though the use of a string of characters, named '*key*'. In that case, the decryption can be done only if you know the key used in encryption.
- *Authentication* is verifying the identity of individuals involved in a communication.
- Verifying the integrity is the test that certifies if the message has not been altered during transmission.
- The *digital signature* is a string derived from the encrypted message that identifies the sender and verifies the integrity of the message.
- A *digital certificate* is a document of identity that allows virtual entities to identify the network.

3.2 Cryptography algorithms

Cryptography provides a set of algorithms and processes to make the message unintelligible. Some of them are very powerful and have resisted to several attacks; other are less safe, but equally important.

The goal of every encryption algorithm is to make as more complicated as possible the decryption of a message without knowledge of the key. If the encryption algorithm is effective, the only way to decrypt the message is to try, one by one, all the possible keys until you find the right one, but this number grows exponentially according to the length of the key. The most delicate operation in a cryptographic system is the generation of the key. To be really effective, it must use keys of considerable length, besides the keys must be generated in a truly random manner so as to be completely unpredictable for a hypothetical decipher system. For this reason, pseudo-random number generators provided by the computer, usually used for games and simulations, should be discarded and more complex systems, which make use of the background noise of the physical world that can in no way be predicted, should be adopted. Good sources of random numbers are the processes of radioactive decay, the background noise in a semiconductor or intervals of time between two actions of the operator in the computer. The most commonly used source exploits the movement of the mouse or the measure in milliseconds of the time of typing.

It appears, however, very difficult to determine the effectiveness of an algorithm. Sometimes, algorithms that seemed very promising turned out to be extremely easy to violate. For this reason it is preferable to rely on those algorithms that seem to resist the longest.

The encryption algorithms are divided in two classes:

- private key algorithms and
- public key algorithms.

The difference between them is that the first ones use the same key for encryption and decryption, while the second ones use two different keys, one public and one private.

3.2.1 Private key algorithms

Private key or symmetric algorithms are the most commonly used methods. They use the same key for encryption and decryption. Both parties know the key used for encryption, the private key or symmetric key, and only they can encrypt and decrypt the message (Figure 1).

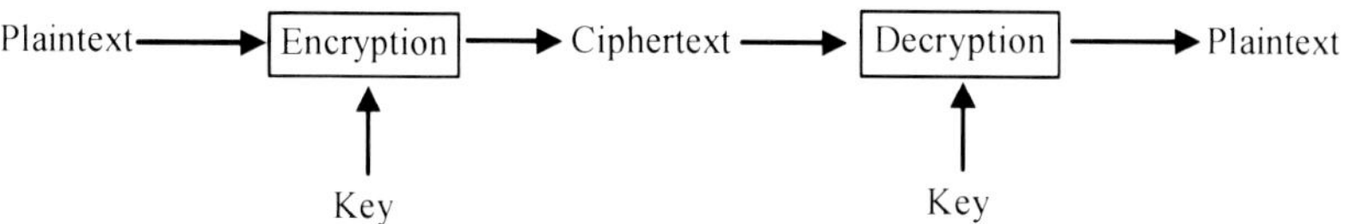

Figure 1: Private key algorithms.

Two types of encryption are available:

- *Stream cipher*: The message is seen as a sequence of bits that are encrypted one bit at a time. They are certainly the fastest but they are considered unsafe, although the security depends on the algorithm used.
- *Block cipher*: The message is divided into blocks of fixed length encrypted a block at a time. Although they are slower than the previous ones, they are considered more secure because each block is encrypted by mixing it with the previous one.

The RC4, developed by RSA Data Security Inc., is an example of stream cipher encryption. It is a very fast algorithm that accepts a variable-length key. His security was not yet well established but, until now, has resisted very well to different types of attacks. It uses a random number generator; the generated number is then applied using the XOR function to the bit sequence.

Block cipher encryption is more and more used by several known algorithms, such as:

- Data Encryption Standard (DES), developed in 1970 by IBM under the name of Data Encryption Algorithm (DEA)), in 1976 became the standard for the US government. It adopts 64-bit blocks and a symmetric key of 56 bits. Given the limited length of the key, DES is easily violable by computers. A variant called Triple-DES or 3DES was recently developed: it encrypts the message three times with many different keys.
- Blowfish, developed in 1993 by Bruce Schneier, is a symmetric block cipher that uses 64-bit blocks and a variable-length key, from 32 bits to 448 bits, making it ideal for both domestic and exportable use. It received great support from the international community and no successful attacks are known so far. It has been used in some well-known systems like Nautilus or PGPhone.
- International Data Encryption Algorithm (IDEA), developed in Switzerland in 1991, operates on 64-bit blocks using a 128-bit key. It is considered very safe although it is not so fast if compared with others.

The private key algorithms have the advantage of being very fast, suitable for encrypting large volumes of data, but has the disadvantage of requiring the distribution of private key to all the recipients. They therefore need to secure an additional channel through which to distribute the key. This contradiction, in the recent past, has set limits to the development of cryptography, up to the birth of public key algorithms.

3.2.2 Public key algorithms

Public key algorithms use two complementary keys, public key and private key, making sure that the private key cannot be inferred from the public key (Figure 2).

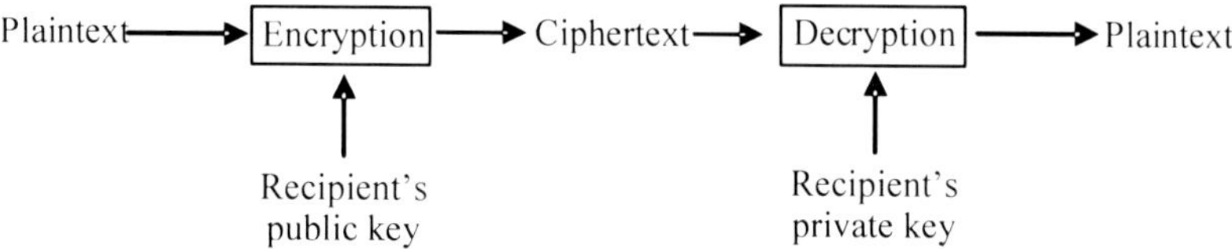

Figure 2: Public key algorithms.

The paradigm of communication is the following: the two partners A and B both have a pair of keys. A requires to B its public key with which A cipher the message and sends the resulting encrypted message to B.

The message encrypted with a public key can be decrypted only with the corresponding private key. Therefore B, using its private key can decrypt the message and read it in complete safety.

With this method only the private key must be kept secret while the public key can be distributed to everyone to send a message to the owner of the key. If an attacker steals this key, he can only encrypt messages without decrypting them.

The most popular public key algorithms are:

- Diffie-Hellman, which is generally considered safe, particularly in the exchange of symmetric keys, when used with a key long enough, preferably at least 1024 bits.
- RSA is the most widely used both for encrypting messages to include the digital signature. It is generally considered safe when used with keys of at least 1024 bits (512 insecure, 768 moderately secure and safe in 1024). It is based on the difficulty of decomposing a number into the product of its first factors. In fact taken two numbers x and y it is very simple to calculate their product $p = x * y$ but it is extremely complicated to decompose p as the product of its components.
- Elliptic Curve, an algorithm that is relatively young, but very slow. It is considered extremely safe, but it has not yet been extensively tested as the RSA.
- Digital Signature Standard (DSS) was adopted by the US government and is used mainly for digital signature.

3.2.3 The technique adopted in practice

Public key algorithms are considerably slower than private key ones, especially in the encryption of great amounts of data. Therefore, in cryptographic systems it is preferred to adopt symmetric algorithms for encryption of messages and public key algorithms for encryption of symmetric keys.

The sender generates a symmetric key, encrypts the message, crypts the generated key with the public key of the receiver and sends both the message and the key. The recipient deciphers the symmetric key with his private key, and finally deciphers the message with the symmetric key.

3.3 Digital signature

Thanks to the complementarity of public and private keys, a string encrypted with a key can only be decrypted with the symmetric one. Thus, deciphering a text using a key ensures that it has been encrypted with the complementary key.

Digital signature algorithms exploit this feature to verify the real origin of the message (sender authentication).

The digital signature is a string, called fingerprint, derived from the message by applying a particular algorithm. The fingerprint is unique, it is encrypted using the private key of the sender and sent along with the message. The public key is the only one that can decode the message (the keys are complementary). The deciphering of the signature using the public key is an evidence that it has been encrypted by the sender or by someone in possession of its private key. Moreover, comparison of the deciphered string with a string made from scratch from the message using the same algorithm allows to check the integrity: if the two strings match, the message is intact [6].

3.4 Hashing algorithms

The digital signature algorithms based on hashing algorithms. They are one-way algorithms that produce, starting from a variable-length string to a string of fixed length (typically between 64 and 255 bits) that is characteristic of the string data.

Their power is due to the following peculiarities: given a string of hash, it is computationally impossible to derive the message from which it was generated and it is computationally impossible that there are two messages that produce the same string of hash, the same algorithm, applied more times to the same message, always produces the same hash value.

The most popular algorithms are the following:

- Message Digest Algorithm 5 (MD5), developed by RSA Data Security Inc. It is the successor of MD2, and MD4, algorithms now in disuse. It produces 128-bit hash from strings of arbitrary length, it is widely used and is considered reasonably safe.
- Secure Hash Algorithm (SHA), developed by National Institute of Standards and Technology (NIST) and National Security Agency (NSA), is used by the US government and produces strings of 160-bit hash from strings of arbitrary length. It is considered quite safe. Usually used in conjunction with the DSS.

3.5 Certification

Digital certificates play a vital role in public key cryptography. Their goal is to authenticate an individual certifying that the public key declared by him, really belongs to the subject for which it was issued. A certificate is, in fact, a digital identity card. Like a real document, it contains a set of attributes that identify the holder of the certificate. It is issued by an entity, called the Certification Authority (CA), officially recognized by society, as the entity that guarantees the authenticity of the information contained therein. Usually, in addition to information relating to the subject, the certificate contains the public key, some information relating to the certification authority which issued it, the digital signature affixed by the certifying authority and the period of validity [7].

The paradigm is the following: an individual completes a certificate request with the data and public key and sends it to a certificate authority. The authority verifies the authenticity of data and, if the response is positive, it produces a certificate that is sent to the applicant signed with the private key of the authority. The applicant may now send the certificate to another individual in order to be authenticated and to give the public key. The verification of the identity is conducted by checking the signature on the certificate from a CA that, therefore, makes available to everyone its public key. Actually it provides its own self-signed certificate or signed by another certificate authority.

3.6 Conclusions on cryptography

Currently, encryption is receiving great attention from the scientific world due to the general development of e-commerce. For this reasons protocols have developed for the safe navigation on the Web, such as the https protocol based on Secure Socket Layer (SSL) protocols to ensure confidentiality of electronic mail, as S/MIME and mail servers that support the SMTP, POP3 and IMAP protocols, over a secure channel using SSL. Moreover, digital signature techniques are increasingly used. The pair certificate–key, sometimes called personal security environment (PSE) will probably represent our future identity card.

4 Bluetooth architecture

There are different levels of security implemented in the Bluetooth architecture, according to the services offered and the involved devices. The link level (baseband) is what gives many of the basic security features. In addition, Bluetooth specifications provide for a security manager handling all the security procedures at the service level.

4.1 Security levels

When two Bluetooth devices connect each other for the first time, both are able to determine if the other one is reliable or not. A reliable (trusted) device have access to all services and it is said to have a fixed relationship. An unreliable (untrusted) device has a temporary relationship and has limited access to services. Regarding the security services there are three levels [Mull]:

- *Services that require authentication and authorization*: access is automatically granted to the trusted device, while the not-trusted devices must pass the authentication procedure.
- *Services that require authentication only*: authorization is not required.
- *Services that are open to all devices*: To ensure that the access is guaranteed and the devices do not require any form of authentication.

The security level of a service, in addition to authentication and authorization, uses an extra attribute: encryption. In this, before access is granted, the link should be sent in encrypted mode. This type of information with regard to the services stored in the database of the security manager. If you have not defined any level of security, then the default one is that one active: it requires authentication and authorization for inbound connections, and only authentication for the outgoing connections. In general, the access granting to a service does not guarantee access to other services on the same device and does not automatically guarantee future or uncontrolled access on the same service on the same device. Bluetooth General Access Profile ranks security in three ways that affect the functionality and applications of a device:

- *Non-secure mode*: no measure of security on the device has been initialized and the safety function of the link level has been bypassed. In this way the device works faster and consumes less energy. This mode is used in applications where security is not strictly necessary.
- *Service-level Security Enforced Mode*: A device starts the security procedure only after the channel choice on the Logical Link Control and Adaptation Protocol (L2CAP). In this mode applications with different security requirements can coexist simultaneously in the run state. This method thus provides flexibility between different types of applications.
- *Link-level Security Enforcement Mode*: The device initializes the security procedures at the lowest level of the protocol, before completing the link-level on the Link Manager Protocol (LMP). Once the channel has been established, a safe physical connection between devices is created. This mode is used in critical applications where security is essential.

The methods for user authentication and for encryption of transmitted information are set in the heart of the security specifications of the Bluetooth.

These methods are implemented in link layer and in this level, to improve security, we use different keys. They also implement four safety parameters:

- Bluetooth Device Address (BD-ADDR) is a unique public address to 48 bit possessed by each device Bluetooth.
- RAND is a 128-bit random number, generated for each transaction, from the same device, which builds the channel and the encryption key. It is extremely unlikely that a value is repeated over time while an authentication key is valid, making impossible to predict its value.
- Key Link is a private key 128-bit used for authentication between two or more units. It can be either a temporary key (a key to initialize or a master key) or semi-temporary (unit or combination key).
- Encryption key is a secret key to 8-bit or 128-bit encryption used each time a session is started.

4.2 Security manager

The security manager is the key component in the Bluetooth architecture (Figure 3). It fact, it manages all the security procedures at the services level, and maintains a database of different security requirements of services and levels of safety device. The security manager handles the following tasks [8–10]:

- *Information storage, related to security, on devices*: the security manager uses the device database to store addresses, the key link and the level of confidence of slave devices that have communicated with the master unit in the past.

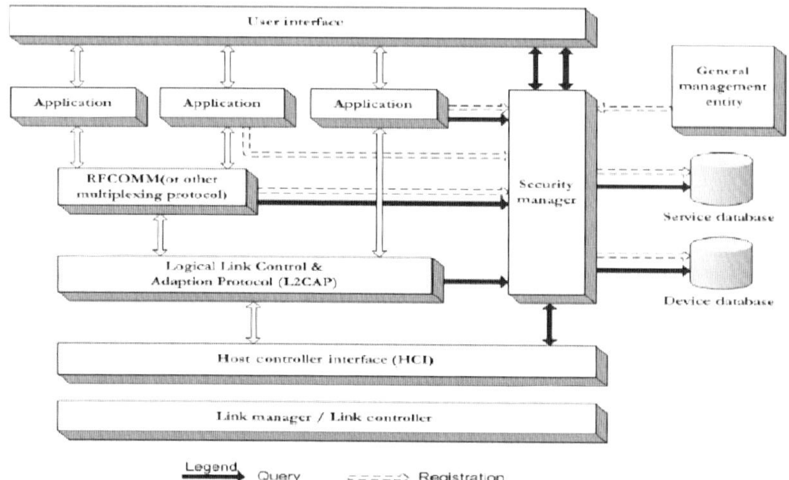

Figure 3: The Bluetooth architecture

- *Information storage, related to security services*: a database of service ranking the applications, before they become available, according to their level of security.
- *Response to requests for access by applications or implementations of the protocol*: each time a requested data unit, or access to a particular service, the security manager will grant or refuse the requested access, basing on the security information of both the service and the unit.
- Imposing the procedures required for encryption and/or authentication before connecting to the application through the stack protocol.
- Initialize or process the input from a user of the device to configure the trust settings of the device.

4.3 Ad Hoc networks

An ad hoc network is a collection of wireless mobile nodes dynamically forming a temporary network without the aid of network infrastructure or centralized administration. All nodes are able to move and can inter-connect in an arbitrary manner. Bluetooth technology, independent of fixed infra-structure, from the beginning has been developed for mobile devices and, as a result, it is able to create ad hoc networks, for this it is perfectly keeping with the philosophy of UC. However, Bluetooth has some different characteristics and properties that do not support some of the routing protocols of ad hoc networks. As a matter of fact, the routing of data between the master and the slave in a small network is a particular safety problem (Piconet). A solution to this problem is to form a combination of keys that can be used to encrypt traffic. The master unit generates a combination of keys with each slave, thereafter data can be sent from a slave to all other slaves of the network, through the master, who knows all the keys. Another approach for the development of a secure ad hoc network is based on the master key. This implementation requires that all devices on the network use the same key to encrypt the traffic, eliminating the need for passage through the master. Another major problem is that Bluetooth is a network link-oriented, where the first operation performed is the creation of connections between devices. On the other hand, the ad hoc networks are networks broadcast-oriented, where the devices continuously emit radio messages and all other units are listening and can process the received signals [11].

Certain types of routing in ad hoc networks require the location information of a particular device. In this case, Bluetooth technology cannot satisfy that request because it may provide information regarding only the signal strength of a device that is communicating with another. This information, however, can be used to provide services. Currently, the routing of ad hoc networking with Bluetooth technology is a topic of research [11].

5 Authentication systems

5.1 RADIUS

Remote authentication dial-in user service (RADIUS) is a system of authentication and account management used by many Internet service providers to authenticate their users. RADIUS is also suitable for wireless authentication. As the name suggests, it was developed to serve remote users through dial-in (e.g. remote access). Radius simply authenticates users. A very important part of RADIUS is its interoperability, which enables the server to communicate with other servers that are based around the same protocol.

5.1.1 Configuring the RADIUS

In client–server mode, the user communicates with the network access server (NAS) and, in turn, the NAS acts as a client on the RADIUS server. The NAS and RADIUS server communicate with each other via a network or a point-to-point protocol. As previously mentioned one of the features of RADIUS is that you can communicate with other servers, based or not on the same protocol. The basic idea is to have a central source of information for authentication. The user is prompted to provide the information to the NAS server for authentication, such as username, password and PPP package for authentication. At this point the client (NAS) has access to RADIUS, creating a request access message and sending it to the RADIUS node (server). This message contains information about the user, which are called attributes. Attributes are defined by the system manager of RADIUS and may therefore vary. They can be password, the user ID, the destination port, the client ID etc. If the attributes contain important information they must be protected through the MD5 algorithm. Additionally, all transactions between the client and the RADIUS server must be authenticated and the password exchanged between the two devices must be encrypted.

5.1.2 Exchanging messages

The server and the client must share a secret key. As the access request to the RADIUS server is received, the server verifies to have the secret key for the client, if this is not the case the request is discarded. Once the initial check is over, the server queries a database that contains information needed to authenticate the user. If authentication is compliant with all requests, the server sends a response to the user as a message to reply. The client in turn can relay the answer to the form of a prompt and, in any case, it must send the message again to request for access, with a few different fields, the main of which is a reply to the encrypted response from the server (Figure 4).

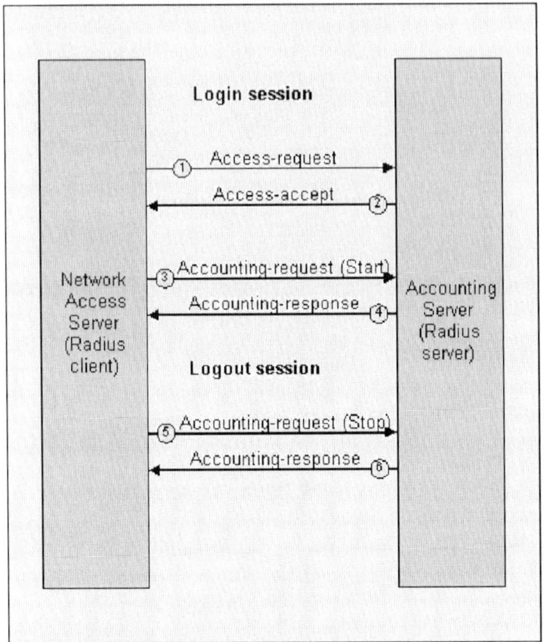

Figure 4: The exchange of RADIUS messages.

To the user it is presented as a random number, which should be encrypted and the result of encryption should be retransmitted to the server. The server receives this message and examines it, if all is ok, it sends a message of access to the client. The RADIUS protocol goes beyond the authentication support, as the granting message contains configuration information, such as PPP, user login and so on. With the RADIUS node all the information necessary to support the session on a network are provided, e.g. an IP address for the session, compression services, the maximum transmission unit (MTU) and so on. The NAS client can support PAP and CHAP protocols. In this case, the NAS sends the client ID and password in the message asking for access (specifically in the user-name and password fields of the message). If using the CHAP protocol, the NAS client generates a response and sends it to the user. According to the rules of the CHAP protocol, the user responds with IP CHAP and CHAP username. At this point the NAS client sends to the RADIUS server the message of access request, containing the CHAP information. The RADIUS server uses the UDP protocol, since if the first authentication to a server fails, it must be conducted on a secondary server. UDP also simplifies the use of multi-threading (where the user request generates several processes to reduce the delay in securing the authentication). The RADIUS, of course, has its limits, residing in the structure of command and the address space of attributes, resulting in poor capability to introduce new services. RADIUS, working on UDP, has no mechanisms of timing or retransmission. For these reasons, manufacturers have implemented several versions of these procedures.

5.2 Kerberos

Developed by the Massachusetts Institute of Technology, Kerberos, whose name takes inspiration from the three-headed dog that in Greek mythology was the guardian of the gates of hell. It deals with the authentication of users, the generation and maintenance of encryption keys. Kerberos is a distributed authentication service that allows a process (client) to authenticate on behalf of a user (user) to a verifier, without sending data across the network that could allow a hacker or a verifier to impersonate the user. Kerberos uses a series of encrypted messages to prove to a verifier that a particular user is working on a client [12].

Kerberos is composed of three elements:

- *Authentication server (AS)*: It deals with storing passwords and interacts with the client workstation to authenticate the user. This interaction also includes the creation and sending of a ticket granting. This is used by the client to obtain a service granting ticket from the ticket-granting server.
- *Ticket granting server (TGS)*: It provides the client a service-granting ticket for receiving services from one application server. This server was introduced to prevent a user to retype his password each time a request for authentication is asked.
- *Application server (AP)*: The application server provides the services desired by sending to the client.

In summary, the Kerberos protocol provides that at first a client contacts the AS to retrieve a ticket valid for the tags: This ticket is called ticket-granting ticket (TGT). When the client invokes the service for the first time, it invokes the TGS to get a session ticket that allows access to the service, then, for the duration of the session, the client directly invokes the service sing the same ticket.

Although the protocol can be considered secure enough, it also has its weaknesses, and there are some peculiarities that we must not underestimate:

- Clients and servers must necessarily maintain safe their secret key. If any attacker obtains the secret key, the system can become vulnerable to spoofing attacks.

- Kerberos is not capable to manage denial of service (DoS) attacks. With a simple DoS attack, you can prevent an application to participate in a process of authentication. Even Password Guessing attacks are not managed by Kerberos.

5.3 Other secure authentication systems

5.3.1 Biometrics: definition and fundamental components

In recent years the idea that the use of biometric techniques can be a good compromise between safety and ease of use has grown.

Biometrics is the science that studies the analysis of those biologic characteristics of the individual, which are unique and unrepeatable, allowing therefore the identification. The biological characteristics are divided, more precisely, in physiological (linked in a steady manner to the anatomy of the human body) and behavioural characteristics (linked to the personal way of life of each human being) that are easier to integrate, but less reliable.

The approach to the techniques in the field of biometric security completely upsets the concept of passwords. There is a migration from authentication based on something that the user knows (password or PIN), or something that the user has (magnetic identification card or smart card) to an authentication type based on what the user is (biological characteristics that biometrics can identify).

The biometric technique offers the best advantages in terms of safety and convenience: personal biologic characteristics cannot be borrowed, cannot be stolen, cannot be forgotten and they are virtually impossible to replicate (the possible use of plastic surgery for replicating the anatomical feature is not taken into account).

There are two different procedures for the use of biometric systems: *verification* and *identification.*

Verification (or authentication) is used when the user is already registered, the user declares the identity and the system acquires the biometric feature. This is compared only with that which you have already saved in a database.

The identification (or search) is used, however, where the identity of the subject is not known a priori. In this case, the extracted biometric feature of the individual is compared with those in a database to establish the identity of the subject. If the feature is not contained in the database, the identification process gives a negative result.

Of course, identification is more expensive than verification; it needs more resources and presents an accuracy that decreases with the size of the database.

The most common physiological characteristics that can be analysed for security are fingerprints, the geometry of the palm of the hand, the retinal or the iris scan and the shape of the face. The most used behavioural features are the analysis of the signature and the speech recognition.

In literature there are various other techniques including body pace examination, scanning the palm, the geometry of the veins of the hand, DNA analysis, body odour and keystroke dynamics on the keyboard.

The process to use a biometric system for security purposes essentially consists of two phases. In a first phase of registration (or enrolment) the biometric characteristic is captured and a biometric template is extracted according to a specific algorithm suitable for the feature.

The template is a mathematical representation of biometric data and is different from individual to individual.

A template can vary the size depending on the amount of information that it contains (to preserve the geometry of a hand a few bytes, for facial recognition will go to several thousand bytes). The template is subsequently stored in an area of memory. This may be a local structure (the point at which the particular user), a central memory (so it comes to biometric template database) or a memory contained within a smart card owned by the user.

The second phase is the actual identification process. A scan of the biometric feature is performed. This scanning is processed using the same algorithms used in the previous phase of registration and the biometric template is extracted. This template is then compared with those ones stored in the database of the registered users.

In this last step a value of similarity found between the two templates (the one that [13] defines as matching score) is computed. This step allows evaluating and modifying the error sensitivity with which the biometric system works.

The identification process can be optionally followed by a phase of adaptation: the system, having recognized the user, updates its biometric template in the database with what has just been calculated in the identification. This is a way to manage the problem of change in time of the biometric feature that is updated at every positive identification of the user.

5.3.2 Hardware keys

A hardware key is just a digital identity card, which must be kept safe. It is an object with memory in it, whose capacity varies from model to model. In this memory, data such as passwords, digital signatures, codes and so on can be written. The hardware security key is mostly based on the reliability of the key holder, as a matter of fact if someone has personal identification data stored on a key and leaves it unattended, then security can be easily compromised.

Apart from that, the degree of security of a hardware key is undoubtedly better to classical password text. However it is possible to 'tamper with' a USB key, reading the contents of memory and thus replicating the key. Obviously it is a procedure that requires specific skills in electronics and the availability of appropriate tools. Yet the skills required and the necessary tools are readily available and anyone with a good basic knowledge of electronics may be capable to perform these procedures.

5.3.3 Smarts cards

There are two types of smart cards available in the market: those with a magnetic band and those with an integrated chip. Among the chip-based cards, the most simple and less secure are certainly the cards with memory chips, while those with a processor can be considered much more secure.

The magnetic bands of cards, should be read with cheap and simple devices and on the Internet hundreds of sites explain how to copy cards using the heads of old VCRs.

Furthermore, the content of the magnetic bands of this type of cards, not only is unprotected, but it is written in clear without any form of encryption. Data present in the band, therefore, can be easily read and also immediately changed.

Chip-based cards have a technology that makes it possible writing data and block them by means of a password. This allows for the protection of the chip contents, both in reading and in writing, and it is possible to encrypt the information written in the memory card with a mechanism of public and private keys.

The card processor resets or alters (as instructed by its creator) the contents of memory if a user tries repeatedly to access memory with the wrong password.

Smart cards can be read by contact (magnetic and chip cards) or by appropriate receivers of radio signals. In this case we speak of contactless cards, i.e. the content is simply read towards a special antenna. These cards have a chip memory in which to insert the data and an antenna to communicate with the player.

Being more complex, magnetic cards have a greater security degree. If on a scale from 1 to 10 magnetic cards are at level 1 (lowest) and the chip cards to level 10 (the higher) contactless cards can be placed in the middle scale (level 2–4).

5.3.4 Proximity tools

Proximity tools are devices which exploit radio signals to perform an authentication process. These systems read the contents of the memory of a special tool available to the person that must be authenticated. The authentication procedure may include local access to an access point (logon to the physical machine), or even to a server on the network. In this case, if the communication channel is secure, the authentication procedure can be efficiently performed.

5.3.5 WAP/UMTS communication as a system of authentication

Wireless Application Protocol (WAP) consists of a set of protocols that allow you to interface programs usually available on cable networks, with relatively limited capabilities that a mobile phone or PDA has compared with those of a traditional computer.

The first standardization of the protocol was made with the WAP 1.0, born in April 1998. Being WAP an open standard, it is imperative that devices, services and applications have at least four requirements:

- *Interoperability*: the terminals of different manufacturers should be compatible with any network.
- *Scalability*: The network operators need to integrate services and applications with maximum flexibility.
- *Efficiency*: the quality of service provided should be adequate to the support network.
- *Safety*: when required is necessary to ensure the integrity and protection of sensitive data that can be captured by third parties.

The WAP architecture is unfortunately not completely adapted to the security model offered by secure connections via SSL for the following two reasons.

The SSL protocol was designed for wired communications type (broad band and a low latency time) involving the personal computer (PC) with high computational and storage capabilities: an SSL transaction with a WAP terminal entail considerable delays in the communication, greatly affecting performance and cost of communications.

The WAP does not provide direct communication between the client and the web server: between them there is always the WAP gateway that acts as a bridge between the terminal and the mobile web server.

To solve the performance problem, a new protocol (named WTLS) has been designed. It is suited especially for mobile terminals and takes into account their physical limitations. This protocol ensures a good level of security greatly reducing the overhead of SSL.

5.3.6 WTLS

The WAP Wireless Transport Layer Security (WTLS) [Wapf] Protocol deals with the security of the WAP architecture. This protocol is derived from the TLS (which is based on the specifications of SSL 3.0), however, it incorporates some new features and is also implemented to run on networks with limited bandwidth and high latency. The WTLS is an optional protocol.

The main goal of WTLS is to provide to a pair of applications the following properties:

- privacy,
- integrity of information and
- authentication.

The WTLS provides these properties by using the same patterns of SSL encryption, but unlike the latter it also works with datagram transport protocols (like UDP).

Such protocols are characterized by the fact that data are transmitted in a fully independent manner and may also be lost, arrive not in order or even duplicated.

These features make the SSL protocol unusable as it is on a datagram transport protocol.

Let us think to the handshake phase between a client and a server: it is unthinkable that it has a success if, for example, the initial request for a secure connection of the client fails to arrive to the server or if the acceptance of a certificate from either of the two entities never get to the other one.

to support datagram a number of mechanisms were then introduced in the WTLS to face the possibility that data do not arrive or arrive in disorder or duplicates are present.

In particular, to overcome the problems above mentioned the WTLS is based on an asymmetrical state machine (i.e. one for the client and another, different, for the server). The interaction of the two machines can synchronize the data that the two entities share on a secure connection. WSTL is based also on the use of time out so in order not to block one of the two entities endless waiting of the response, the control of the validity of the number of sequence of incoming packages and on the merging of a number of handshake messages travelling in same direction in a single package to be sent.

6 Weaknesses and attack methods

6.1 Deliberate attacks

In considering a deliberate attack, it is convenient to distinguish attached component and the technique used by the intruder. A systematic approach identifies all the components of the system both physical (computers, routers, AP and Palm) and logical (files, processes, etc.) and for each of them, it finds all the applicable attack techniques. The result of this approach may be conveniently summarized in a matrix having the components on an axis and the techniques of attack on the other. A cell of this matrix makes it possible to describe if and how a certain technique can be used to attack a certain component.

The attacks on the physical level are mainly aimed at removing or damaging the resources. The main types of physical attack are theft (it is an attack on the availability and confidentiality) and damage (attack on the availability and integrity).

The logical level attacks are mainly designed to steal information or degrade the system operation. To characterize the possible attacks on the security of a system, it is convenient to consider as in any system there is a flow of information from a source to a destination. Then there are four possible types of attacks (Figure 5):

- *Interruption.* Part of the system is destroyed or becomes non-usable. This is an attack on the availability of the system.
- *Eavesdropping.* An unauthorized person obtains access to a component of the system. This is an attack on confidentiality. Interception attacks (and those of changes described in section below) may request a pre-emptive attack on the physical level to install pirate devices to engage the network and to install software to intercept data. The techniques commonly used are based on:
  - ➢ analysis of traffic on the network (local or geographical);
  - ➢ application of analysis of network traffic (sniffing);

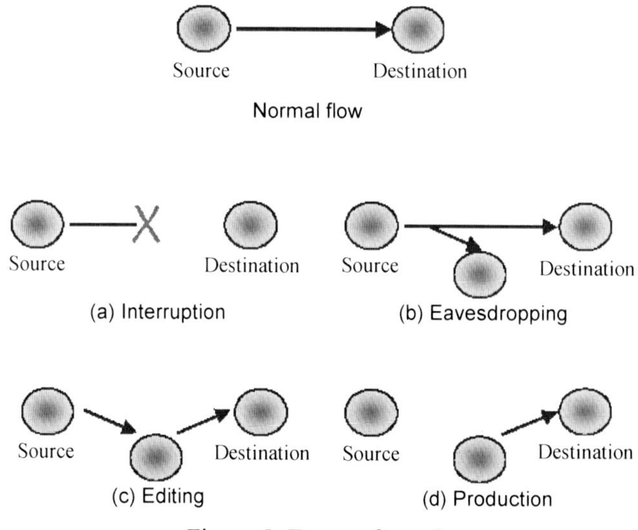

Figure 5: Types of attack.

- pirate server that attacks some routers and assumes the identity of the original server, this attack is based on changing the routing tables of a router (spoofing);
- programs that emulate the services of the system recording at the same time confidential information entered by the user (for example the login program can be emulated when the user enters the username and password to obtain the user's password (password cracking)).

Eavesdropping attacks can exploit inherent weaknesses of protocols and network software or unaware operating system configurations. Interception attacks can exploit the fact that a user has violated any standard of behaviour required by the security policy (such as writing the password under the keyboard). In fact when the system does not provide advanced tools for the user authentication (hardware key, fingerprint reader, etc.), the more frequent intrusion attacks are given by an illegal password.

- *Editing.* An unauthorized person comes into possession of a component of the system, modifies it and introduces it back into the system. This is an integrity attack.
- *Production.* An unauthorized person manufactures new components and places them into the system. Attacks that use these techniques are not designed to access information and services, but simply to degrade the operating conditions of the system. They are considered sabotage acts, and typically they threaten the integrity and availability, more rarely (and indirectly) confidentiality. There are various techniques for disturbing:
  - attacks by viruses;
  - attacks by worms;
  - attacks of 'DoS' type: this is a family of techniques designed to ensure that the system denies access to information and services to duly authorized users. Attacks that use these techniques then threaten the availability requirements of the system. Two typical DoS techniques consist for example in the paralyzing the traffic on the network by generating false error messages or clogging it with specifically generated disturbing traffic.

6.2 Sniffing

Sniffing could be classified as a passive attack to privacy. In short it is like eavesdropping on the door of our neighbour. But in the case of sniffing the consequences for the victim can be much heavier. Being able to intercept packets that transit through a communication channel, it is possible to know private passwords of the victim, which could allow to have the total availability of resources, or even, in some cases, allow the access to systems to take their exclusive control. Let us think for example to the so-called computerized home, when an attacker manages to gain access to our authenticated information, he has at its disposal all the resources of our home, the refrigerator, the heater, etc. Just think like a sniff of action would be extremely damaging at a meeting between co-workers who, with their handheld devices, in an ad hoc connection, exchange data or office information. This would irreparably compromise the firm security. Sniffing is done by means of antennas that intercept radio waves emitted by devices that communicate with each other, and which, when interfaced, display on a screen the contents of the communication packages.

6.3 Denial of service attack

Often system administrators are concerned about attacks on their network that could compromise the integrity or confidentiality of the data of their computers, but not all attacks are brought to get access to a system. Hackers may also execute a much easier attack called DoS. In general, a DoS attack is designed to consume all the resources of the attacked system, preventing other users to use them; therefore we are talking about an attack that limits the availability of a service.

On Microsoft, one of the first programs to create a DoS was the famous WinNuke, a utility able to exploit a weakness of the Windows machine: it suffices simply opening a connection to port 139 and sending a special package, causing the famous blue screen Windows error and a reboot (the port is the access point to a particular network service in a machine, in this case the NetBIOS). The DoS is also a feared attack by PDAs, by which it is possible to avoid the synchronization between Palm and PC through the network. A similar attack is the kiss of death (KoD) that stops the connectivity of the handheld.

6.4 Distributed denial of service

In DDoS attacks the system denies access to information and services to duly authorized users. As an example in early 2000 an attack of this kind paralyzed for several hours some important US sites, making it impossible the use of Internet to millions of users. The principle on which the attack is based is simple: flooding of requests to some random sites, so that they can no longer bear the load of requests and as a consequence they stop working, the system is not available for a long time because all the false requests have to be satisfied. To be able to orchestrate an attack of this kind, it is used

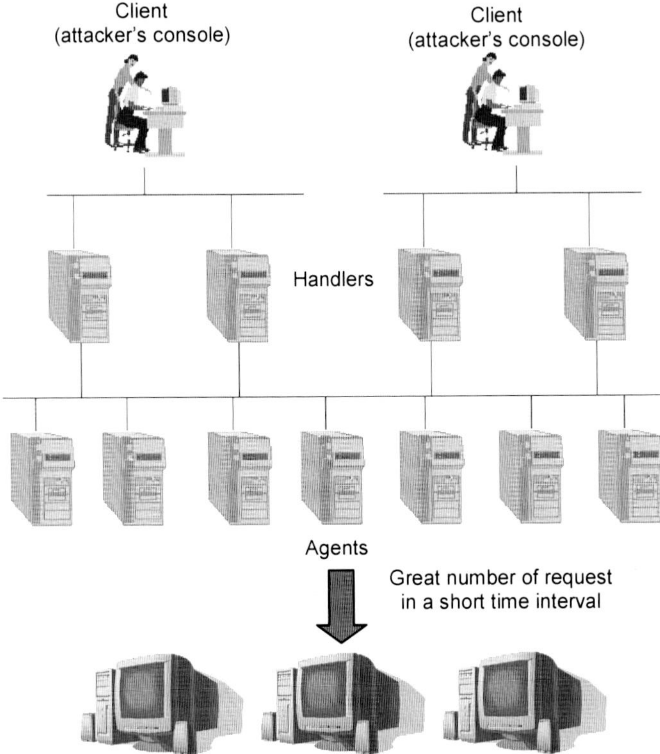

Figure 6: Structure of a DDoS attack.

a distributed mechanism, i.e. a very high number of programs that attackers use, distributed on different computers and devices (often users do not suspect to become 'tools' of the attackers) and synchronized by a few centres. The architecture used was presumably of the type shown in Figure 6. The agents are the programs that actually send random requests to the attacked sites. These programs can be found on thousands of infected computers, meaning that normal users do not know that their machine is running a secret program of this type. The handler programs are the coordinators of agents: a few hundred of these are used to control the attack. These programs are secretly running on infected computers or on some machines of the attackers. Finally, very few client programs are those of the real attackers, the console through which starting the attack by sending appropriate messages to the program handler, which in turn activate the agents. Since each handler can typically manage up to 1024 agents and a client handler up to 1024, it is easy to understand how the power of this structure is impressive and can be controlled by very few attackers.

The novelty of this scheme is that the attack takes place in two phases: the first (lasting months) in which a large number of computers (primary victim) become infected, covering the roles of agents and handlers. The infection mechanisms can be as classic viruses, or code injection.

The second phase is the DoS attack itself, during which, in a few moments, the primary victims are used to send the requests and packets to the targets of the attack, the so-called secondary or final victims. An attack to a centralized authentication server, for example, can block the use of many services required by users. The mechanisms for authentication for UC are oriented to a decentralization that certainly lowers the security threatening for users to attacks of this kind.

6.5 Sleep deprivation torture

The most interesting type of DoS attacks are those that take into account the link between security and preservation of power. Power supply of devices such as PDAs, just because of their mobile nature, is granted by batteries, which have always limited life. If a PDA has little energy in batteries, it tries, whenever it can, to go into stand-by mode to preserve energy as much as possible. This particular attack, efficient and selective, tries to keep 'awake' the PDA until the battery is discharged. As a result you get of course the momentary removal of the device [STAJA00]. But the attacker might also want to isolate a device, disabling all those with whom it communicates, creating the so-called ring of evil.

6.6 MAC address spoofing

The phrase 'MAC Address spoofing' in this context refers to a change in the MAC Address of the attacker to any other value. MAC spoofing is conceptually different from traditional IP Address spoofing, where an attacker sends data from an arbitrary source and simulates another IP address. Almost all 802.11 cards in use allow the alteration of their MAC addresses. For example, using the open source drivers for Linux, a user can change their MAC Address with the ifconfig tool, or with a small C program that suitably calls the function ioctl(). Windows users are allowed to change the MAC address by selecting the properties of their network adapter driver in the control panel.

An attacker may choose to alter its MAC Address for several reasons, including confusing its presence on the network, excluding the listings of access control or impersonate an already user authenticated:

- *Presence hiding*: one can choose to change his MAC Address trying to elude intrusion detection systems of the network (NIDS). A common example is that an attacker runs a script to attack with a random MAC

Address for each connection attempt. Such an attack would pass unnoticed to the applications for the analysis of network activities, such as NetFlow, which indicates the upper-level network or large amounts of traffic from a single source address.

- *Bypassing the access control lists*: These lists are used as the basic form of access control on WLANs, administrators typically have the ability to configure the access point to only allow registered MAC addresses to communicate on the network. An attacker could get around this form of access control passively monitoring the network and compiling a list of MAC addresses that are allowed to communicate. With the list of authorized MAC addresses, an attacker is free to declare your MAC Address with an authorized address, excluding the security mechanisms.
- *Impersonation of an authorized user*: Certain security authentication devices of WLAN hardware rely on authentication credentials to the MAC Address of the client source. After a user has been authenticated, the Security Gateway allows only traffic based on a dynamic list of allowed MAC addresses. An attacker who wishes to avoid the security of the system only needs to check the activities of the network to obtain an authorized MAC Address, and then alter its own MAC Address to take the identity of the user authenticated earlier.

6.7 Attacks on Smart Cards

Being able to conquer the secrets stored in a Smart Card is not simple. Indeed its nature makes it one of the safest ways to protect users. Attacks on Smart Cards can be categorized into two distinct families: *invasive* attacks and *non-invasive* attacks.

Invasive attacks require hours in specialized laboratories, with very expensive machines. A Microprobing action gives access to the chip surface and integrated circuits are directly altered [14]. A type of attack of this kind is irreversible, i.e. hardware is definitely compromised. In addition, the owner of the card will easily notice the attack and can immediately revoke the keys.

In contrast the non-invasive attacks are very difficult to identify. They can be divided into three types:

- *Software*: the attacks are more common and certainly the least expensive. They exploit the vulnerability of protocols, cryptographic algorithms, etc.
- *Eavesdropping*: The attack is aimed at analysing all the connections and CPU electromagnetic radiation.
- *Fault generation*: the goal is to induce a malfunction in the processor.

7 Security on wireless channels

7.1 Bluetooth

Bluetooth can be considered quite safe, but it still shows some specific weaknesses in security. First, it is plausible stealing the keys of the key ring and the encryption keys, thus having the opportunity to 'spy' (eavesdropping) and to impersonate the victim device. Another form of Bluetooth attack has highlighted that the benefits of technology may become disadvantageous for safety. In fact, many Bluetooth devices can record the movements of another device that uses the same technology, without the owner being aware of it.

7.1.1 Eavesdropping and impersonation

The theft of the keys can be done passively eavesdropping traffic, participating actively in a broadcast or carrying a 'man-in-the-middle' attack [15].

In the first case, the attacker 'guesses' all the PINs of a given length, and the accuracy of each PIN is verified by conducting the second operation of the initialization protocol. This verification is based on assumptions of the attacker and on random strings that are exchanged between the devices. Note that this type of attack is done offline and that the attacker receives the data in an entirely passive manner. The PIN length can vary between 8 and 128 bits, but in most cases a common format of 16 bits (4 digits decimal) is used. Also, if there is no PIN available, a default value of zero is used. This makes the PIN Crunching immediate [16].

Active theft needs to be made on the first initialization step of the protocol, dedicated to key generation, performed using a PIN (to guess). The second step of the protocol is carried out with the victim device, with which the attacker has done the first step of the challenge-response protocol. If the conjecture that the PIN is correct then the victim will give an answer like 'Correct'. If the PIN is not guessed, then the attacker can obtain the challenge-response script that can be used offline to try all PINs that the attacker would try.

In detail a key is calculated for each initialization PIN, then the key verification algorithm and the obtained script are tested on the victim. At this point when the algorithm displays 'Correct', the attacker has obtained the PIN of the device belonging to the victim. Using the method of active theft the attacker has more time for the off-line PIN-Crunching. After the key initialization has been obtained, all other keys are accessible to the attacker [17].

In the third case, the attacker assumes the identity of the two devices that communicate each other, using a type of man-in-the-middle attack. If the attacker is in possession of the key link for the two external devices, then the attacker can communicate with both the victims by making believe to each one of the two of being the other one [17].

7.1.2 Location attacks

This type of attack is simple and requires no theft of the key. A Bluetooth-compatible device can be in two states: detectable or not detectable.

The former allows a device being 'seen' from any other; the latter prevents all communications to or from that device. If a device is discoverable, it answers to inquiries made by other devices. If four or more devices intercept the same victim, then, they can coordinate their signal strength to determine the position of the victim. This is possible because all Bluetooth devices transmit their identity to every request [18]. The attack is much more effective to locate the position of a victim, compared with that achieved through a network of mobile phones, because the cells are extended for several kilometers [17].

7.2 WLANs

WLAN based on the 802.11x, for its constitution, make the process of identifying the wireless networks relatively easy [19].

Scanner allows WLAN users to identify a network through the use of a Wireless Network Interface Card, just call NIC, and specific software that searches for access points. As long as there was a limited number of applications for scanning of WLAN, NetStumbler was the most popular applications for Windows platforms, not only because it was and is free, but also because of its simple graphical interface, and its ability to use GPS systems for the detection of the longitude and latitude of an AP. This function is very useful for attackers who want to return later in the region for sniffing the traffic passing on that AP [20].

NetStumbler was created by Marius Milner and was spread by a set of his 'followers' [21]. NetStrubler.org [22] is pursuing a project that allows users to insert into the database of the web site the results of their war-driving. Thanks to the GPS function the site has built a map of all APs within the territory of the United States. The results were prepared on a map and you can click on a particular AP and see where it is installed with precision. Having a site that identifies the insecure networks of companies and makes known to the whole world is like having a section in the newspaper we read every day where we can find the list of companies that decide to leave open their doors during the night. Fortunately, Netstrubler.org administrators give the ability to 'victims organizations' to request the removal from the list of information on their AP.

The loyal fans of Linux will approach to Kismet [23]. It does not have a user-friendly graphical interface as Netstumbler, and it is not easy to use, but it certainly provides much more powerful functionalities. Kismet is not just a scanner, but also acts as a Sniffer. During the AP searching procedure, the packets can be stored for further analysis. The logging feature allows to store separately, depending on the type of traffic analysis, the captured packets. Kismet, in fact, can store encrypted packages that use 'weak keys', separately and then submit them to a WEP Key Cracker.

At best, it takes several hours to obtain the WEP encryption key, but an attacker in a few minutes identifies an unsecured network. Once a WLAN has been found where WEP protocol is not enabled, the traffic can be immediately sniffed. If the target is a free access to the network, the attacker only needs to obtain a valid IP address, a reachable goal through the use of a DHCP on the WLAN.

7.2.1 Breaking WEP keys

In this section we discuss the evidence that gave J. Ioannids and A. Rubin of AT&T that the WEP suffers from a serious weakness that can be easily used to decode the traffic in a wireless network. Implementing their attack, Ioannids and Rubin had set out to achieve three goals:

- The hardware and software necessary for carrying out the attack should not have been very expensive.
- Demonstrate that the attack can be carried out by anyone.
- Optimize the algorithm to break the WEP key that was previously used by Fluhrer, Mantin and Shamir [24].

Achieving the first objective was not very difficult, because the biggest expense is due to face was that of the wireless network adapter. The software they used was based on the utilities included in the Linux operating system, even if the network can be found for free on most systems like Windows and MacOS.

To implement the attack we have to search those initialization vectors (IV) that have the key setup for the algorithm in a state (S) that contains information about the key. If the package checks this condition we will refer to it as resolved. It is relatively easy to control when a particular package provides an initialization vector and the resulting output byte verifies the resolved condition. Each resolved package hide only information about a key byte, then all the key bytes must be correctly guessed before any package gives key information on last byte. We use the word 'guess' because the attack has a statistic nature. Every package that meets the resolved condition gives a rate of 5% to guess the correct key byte and 95% for making a mistake. However, observing a number of these resolved cases, it is possible to get more and more closer to the real key bytes.

Through this algorithm it is necessary to capture from 4,000,000 to 6,000,000 packets to decrypt a WEP key from 128 bits. The two attackers were able to decrypt a password to 129-bit after two days, but in networks with high traffic, it can be assumed to complete the attack in less than a day.

The initial algorithm used by Fluhrer, may be changed to increase the performance of the attack to recover the WEP key. It should be noted that these changes do not undermine the successfulness of the attack, and they can greatly reduce both the time and memory space required for an attacker. These changes affect the choice of different IVs processed in parallel instead of a single type, exploiting the availability of a few keys in WEP implementations and considering particular cases of packages that verify the resolved condition. This new algorithm allows decrypting a WEP key from 128 bits rather than capturing 5,000,000 packets but only 1,000,000, so the success of the attack is ensured within hours. This type of attack is completely passive and therefore not detectable.

There are a large variety of tools for cracking WEP, which does not require any technical knowledge on the WEP protocol and its functioning; WEPCrack and AirSnort are two of the most popular. WEPCrack is a set of Perl scripts designed to decrypt WEP keys, using data collected by the sniffer. AirSnort, on the other hand, includes both features. It gets the traffic it needed to break the key without the help of an auxiliary sniffer.

7.2.2 AirSnort

AirSnort is a Linux-based tool written by Blake Hegerle and Jeremy Bruestle to exploit the vulnerabilities of WEP [25]. One of the difficulties of auditing with the use of applications that we are describing is the fact that not all of them are compatible with the same wireless cards. The compatibility is low due to lack of availability of drivers for the cards. If you want to use these tools, we are faced with a difficult problem that can be temporarily solved with the purchase of at least two wireless network cards. NetStumbler and many other Windows-based applications, require a NIC that uses the Hermes chipset, while Airsnort and many Linux-based applications are compatible only with cards that use the Prism2 chipset (AirSnort 2.0 also requires a support ORiNOCO cards with appropriate patches for the orinoco_cs driver).

Various tests and trials have been conducted before the right combination of Linux kernel, PCMCIA cards, wlan-ng drivers and versions of AirSnort, began to give acceptable results.

Once AirSnort is running, the NIC should be in 'random mode' (promiscuous mode) and set the correct channel to find the Wlan. This channel is derived from the scanner of WLAN, previously used to locate the WLAN. AirSnort run a shell script (dopromisc.sh) that automatically launches the NIC in promiscuous mode, with the right channel configured. AirSnort includes two separate applications: capture and crack.

Airsnort also displays the number of 'interesting packets' (also known as weak keys) captured. AirSnort is efficient because it does not capture all packets that are encrypted, but only those that also serve to break the WEP encryption key. The packages affected are those that in the second byte of the IV have 0xFF.

If the number of packets obtained, containing the required information is sufficient, the application will return the shared WEP key. The failure of the attempt to break the keys does not affect the capture process. According to the read-me file AirSnort, 1500 packets are sufficient to be able to decode a 128-bit key. The processing time depends on both the key size and traffic on the network. When network traffic is close to 11 Mbps, encoding WEP key to 40 bit can take 3–4 hours. These times are obtained under optimal conditions, but certainly show that WEP can be bypassed and an attacker only needs 'a little patience and a little time' to access data.

Although it is still efficient, the AirSnort project is no longer maintained, being replaced by AirCrack-ng [26].

7.2.3 WEPCrack

WEPCrack [27] is a project of SourceForge [28], managed by Paul Danckaert and Anton Rager. This application is easier to use than AirSnort. WEPCrack is a set of Perl scripts and requires no configuration. However WEPCrack is used in conjunction with an external sniffer, since it has no catch function for traffic.

The process of capturing data must be completed before using WEPCrack. To capture the data, prismdump is a valuable support. It is a command line sniffer that does not require any argument and simply captures all traffic recognizing all headers 802.11x, which, of course, are essential for capturing the WEP traffic. For functions of traffic interception, it is based on the libraries included in the Ethereal protocol analyzer [29]. Once a sufficient number of encrypted data has been captured, the weak IV and the first byte of encrypted data must be extracted to a separate file. When a sufficient number of data has been extracted, WEPCrack can start.

WEPCrack is certainly less efficient than AirSnort, because it does not capture data that are used for decoding, requiring the user to extract data of interest.

Acknowledgements

This chapter was written with the contribution of the following students who attended the lessons of 'Grids and Pervasive Systems' at the faculty of

Engineering in the University of Palermo, Italy: Bono, Carlini, Castellucci, Cavaliere, Cimò, Colomba, Cutini, Giglia, Greco, Mauro, Nicotra, Panzavecchia and Patti.

Authors would also like to thank Giovanni Pilato for his help in the Italian–English translation.

References

[1] Skow, E., Kong, J., Phan, T., Cheng, F., Guy, R., Bagrodia, R., Gerla, M. & Lu, S., A security architecture for application session handoff. *Proc. IEEE Int. Conf. on Communications ICC*, **4**, pp. 2058–2063, 2002.

[2] Stajano, F. & Anderson, R., The resurrecting duckling: security issues for ubiquitous computing, *IEEE Computer*, **35(4)**, pp. 22–26, 2002.

[3] Summers, R.C., *Secure Computing: Threats and Safeguards*, McGraw-Hill: Hightstown, NJ, 1997.

[4] Stallings, W., *Network Security Essentials: Application and Standards*, Pearson Education, New York, 2nd edn, 2007.

[5] McGrath, R.E., *Discovery and Its Discontents: Discovery Protocols for Ubiquitous Computing*, University of Illinois at Urbana-Champaign, Champaign, IL, Technical Report UIUCDCS-R-99-2132, April 2000.

[6] Merkle, R. A certified digital signature, *Advances in Cryptology – CRYPTO '89, Lecture Notes in Computer Science*, ed. G. Brassard, Springer-Verlag: Santa Barbara, CA, Vol. 435, pp. 218–238, 1990.

[7] Canetti, R., Universally composable signature, certification, and authentication. *Proc. 17th IEEE Computer Security Foundations Workshop, CSFW*, pp. 219–233, 2004.

[8] Muller, T., *Bluetooth Security Architecture: Version 1.0*, Bluetooth White Paper, Document # 1.C.116/1.0, July 15, 1999.

[9] Janssens, S., *Preliminary Study: Bluetooth Security*, January 2005, http://drdeath.myftp.org:881/books/Bluetooth_security.pdf, retrieved on June 2009.

[10] Sun, J., Howie, D., Koivisto, A. & Sauvola, J., *Design, Implementation, and Evaluation of Bluetooth Security*, www.mediateam.oulu.fi/publications/pdf/87.pdf, retrieved on June 2009.

[11] Willekens, J.P.F., Ad-hoc routing in Bluetooth, *Lecture Notes in Computer Science*, **2213**, pp. 130–144, 2001.

[12] Steiner, J.G., Clifford Neuman, B. & Schiller, J.I., Kerberos: an authentication service for open network systems, *Proc. Winter 1988 Usenix Conference*, February 1988.

[13] Silverman, M. & Liu, S., A practical guide to biometric security technology, *IEEE IT Professional*, **3(1)**, pp. 27–32, 2001.

[14] Matthews, A., Side-channel attacks on smartcards, *Network Security*, **2006(12)**, pp. 18–20, 2006.

[15] Yaniv Shaked, A., *Cracking the Bluetooth PIN*, School of Electrical Engineering Systems, Tel Aviv University, http://www.eng.tau.ac.il/~yash/shaked-wool-mobisys05/, retrieved on June 2009.
[16] Vainio, J.T., *Bluetooth Security*, Helsinki University of Technology, http://www.iki.fi/jiitv/bluesec.pdf, retrieved on June 2009.
[17] Jakobsson, M. & Wetzel, S., Security weaknesses in Bluetooth. *Proc. 2001 Conf. on Topics in Cryptology: the Cryptographer's Track At RSA*, April 08–12, ed. D. Naccache, Lecture Notes in Computer Science, vol. 2020. Springer-Verlag: London, pp. 176–191, 2001.
[18] Xydis, T.G. & Blake-Wilson, S., *Security Comparison: Bluetooth Communications vs. 802.11*, Bluetooth Security Expert Group, http://merlot.usc.edu/cs530-s05/papers/Xydis02a.pdf, retrieved on June 2009.
[19] Stanley, R.A., Wireless LAN risks and vulnerabilities, *Information Systems Control Journal*, **2**, 2002.
[20] Borisov, N., Goldberg, I. & Wagner, D. Intercepting mobile communications: the insecurity of 802.11, *Proc. 7th Ann. Int. Conf. Mobile Computing and Networking*, Rome, Italy. MobiCom '01. ACM: New York, NY, pp. 180–189, 2001. DOI: 10.1145/381677.381695.
[21] http://www.stumbler.net/
[22] http://www.netstumbler.org
[23] http://www.kismetwireless.net
[24] Fluhrer, S.R., Mantin, I. & Shamir, A., Weaknesses in the key scheduling algorithm of RC4, *Revised Papers From the 8th Annual international Workshop on Selected Areas in Cryptography*, August 16–17, eds. S. Vaudenay & A. M. Youssef, Lecture Notes in Computer Science, vol. 2259. Springer-Verlag: London, pp. 1–24, 2001.
[25] http://airsnort.shmoo.com/
[26] http://www.aircrack-ng.org/doku.php
[27] WEPCrack, an 802.11 key breaker, http://wepcrack.sourceforge.net/
[28] WEPCrack at SourceForge.Net, http://souceforge.net/projects/wepcrack
[29] Ethereal: a network protocol analyzer, http://www.ethereal.com

Chapter 8

Service discovery

1 Introduction

Making a ubiquitous system means to provide devices capable of obtaining information from the environment in which they are incorporated and meet the demands of their clients.

Clients are moving in a random manner into the environment, using devices such as PDAs, to 'discover' the services that the environment offers. But what exactly is meant by 'service'? A service is an entity that can be used by a person, a program or even by another service. A service can be computational capability, memory, a communication channel, a hardware device or another user [1].

In a nutshell, if we want to indicate two possible types of services, we may say, print a document and convert a document from one format to another. To enable future communication between different entities, it is essential that each one of them is aware of what is available in the vicinity, that is, capable of identifying other entities in the vicinity and interact with one of them using any of the services it offers.

The resolution of this problem is called 'service discovery'. All this must be done taking into account the interests of the client, which may be: finding the closest device with shorter queues and in a manner entirely independent of the device's position [2].

However, allowing a user to take advantage of the services provided by the environment is not an easy problem to solve: we must take into account issues such as the position of the user (which may vary) or of other objects and knowledge of place [3]. We will discuss these issues in the next few paragraphs, gradually putting together the 'pieces' to create a system oriented to the discovery of a service.

1.1 Data transmission in ubiquitous systems

This paragraph will explain the techniques used to enable the transmission of data in distributed environments for ubiquitous computing. The interaction plays a crucial role for ubiquitous devices: the reduced potential of the individual computer is balanced by the cooperation between them. Note that this last point is a 'conditio sine qua non' for the ubiquitous system. Ubiquitous devices exchange among themselves a certain flow of information, and this exchange depends on the quality of service provided by the system. The information must be processed, routed and exchanged as quickly as possible (each node has the data needed by a number of other nodes). This goal can be achieved by exploiting the fact that each node of a ubiquitous system is actually a computer but within the limits imposed by its structure. Summing-up, since the tasks of the ubiquitous node are not so complex, it will be sufficient to provide it with resources that, though not being 'excellent', are perfectly suited to the objective (or objectives) that it must pursue [4].

1.2 Objectives

The basic problem in the transmission of technical data in a ubiquitous environment is to determine an efficient plan (scheduling) to determine the order in which to meet in time the required data as soon as they are presented to a node. Until now, one of the most used policies for this regard was the one based on the RxW algorithm. This algorithm calculates for each data the object requested to the node, the product of computer-users (R) who requires it and for the maximum allowed waiting time (W), the time elapsed since the data was required for the first time. A system based on a RxW policy then sends a data pack for two reasons: because it is highly required (high R value) or because some node has required it for a long time. This algorithm responds in a balanced way both to the large demands of a certain feature (hot requests) and to those who are not very (cold calls) so requested. Other proposed algorithms were valid application for environments characterized by applications with a variable size of data requirement, such as the MAX algorithm, proposed by Acharya and Muthukrishnan, whose characteristic was to first meet the demands for larger data [5].

Finally, we mention in this section the fact that the increasingly urgent need to reduce the response time of the node which require information has led to the use of particular components, such as the cache, to speed up the supply of data. As a consequence, the problem of implementing algorithms that support the introduction of such hardware arises.

1.3 Model of ubiquitous node server

Ubiquitous computing, as mentioned, has caused a growth in the development of algorithms for scheduling. All ubiquitous nodes assume that the data required and sent are immediately available to the transmitter. This assumption ignores the fact that the requested data should be taken from secondary data repositories before being sent to other nodes. However, we consider that these data are directly accessible to the node that has been requested.

The flow diagram in Figure 1 summarizes the structure of a generic ubiquitous node which concerns the information exchange. The requests of client nodes arrive at very small intervals of time (obvious consequence of the fact that ubiquitous systems are intended to be used by the mass), and since these requests cannot be instantaneously satisfied, they are queued. We will call it as 'queue of the ubiquitous node', where all the pending requests made by user nodes are placed. The information desired by the mass of customers are typically contained in a disc or alternatively, as often happens, in a cache memory in order to optimize the response speed of the system. Once read from the (cache or mass) memory, data are instantaneously sent to a transmitter, as explained earlier.

Initially, for simplicity, we assume that the communication channel has an infinite speed of transmission, this means that the transmitter does not store data to its input, and as soon as an object arrives, it is immediately transferred to the user who had requested it. In the second stage, we will consider the more realistic case where the channel introduces transmission delays, which is equivalent to non-zero response times in transmission, and therefore, the creation of queues in the input of the transmitter. We will see later the weight on the overall performance of the system in the transmission queues. In any case, the data sent through the communication channel reach the clients who have requested them. (An object data is sent simultaneously to a multitude of users.)

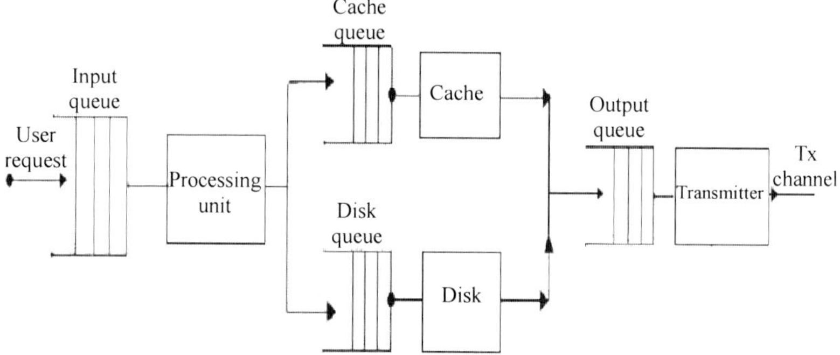

Figure 1: Architecture of a server node.

It should be noted that the number of customers with which the system may interact (i.e. who may require information) may be a sensitive variation in time; for this reason, it makes sense evaluating what happens in the examined scheduling algorithms as the 'number of requests per second' varies. We note that customers use two independent channels to communicate with the server node: one channel for sending and one receiving data. It has been observed that the same object can be simultaneously requested by one or more users: the queues of the server node that receive the requests must take this factor into account to avoid redundancy problems; it means that the queues that are formed in the input to the server are multi-request queues. More precisely, the multi-request queues are formed by groups of common and contemporary requests (we can have two or more multi-requests concerning the same object, but they can be at different points of the queue, i.e. they are present as input to the server at different times). From now on, every time when we refer to a 'request' we will consider a 'multi-request'. Besides, let us assume that the required objects are all of the same size. Moreover, the assumption that the client, after submitting a request, continuously monitors the transmission channel (which from their point of view is the receiver) waiting for a response will be of crucial importance. This means that the server, during the transmission phase, should not wait for the client to connect to the channel to capture the required flow of information. Finally, it will be assumed that there are no errors in the communication channel and we will ignore the propagation delays on the same channel: the delays are considered small compared with the latency that characterizes the response by the server during transmission.

The scheduler of the server also acts on the basis of the actions of the scheduler of the disk and the decisions of the cache manager. Considering this situation, in the following sections we will illustrate four scheduling algorithms for the server, with a single algorithm for the management of disk and some of the most widely used cache management policies.

2 Disk and server scheduling algorithms

Scheduling algorithms on which basis the server decides the order in which to satisfy the data requests are essentially derived by the RxW algorithm [6]. It is suited to satisfy the special needs of data transmission in ubiquitous environments, even with some variations.

With reference to a disk type of memory, we consider the C-LOOK algorithm, which collects sets of data to be considered in ascending order with respect to their position in the cylindrical disk (i.e. about the coordinate radius).

The head starts from the inner part of the disk. On satisfying the requests regarding data in the inner part, the algorithm searches for data requests in the most external part of the disk, then the cycle begins again. (This technique exploits the fact that the rotational latency time is shorter than that of positioning: radial movements are in fact reduced with the technique described.)

Let us consider, first, the absence of a memory cache in the server node, and let us assume that every data request should be met by seeking it on secondary memory. This is done for two reasons: to evaluate the impact that each disc has on server performance and as a consequence of the fact that in some applications the use of a cache might actually not take place. Besides, in some system configurations, the cache size could be so small that it will have a negligible weight on the global characteristics of the node server. For these reasons we will study later the impact of the cache.

The scheduling mechanisms we illustrate here are divided into two categories:

- those using different scheduling algorithms for the memory and the secondary server,
- those that are based on a single scheduling criterion that takes decisions on the basis of information gathered from the queue waiting in the input to the server and those provided by the disk (e.g. location of data on the disk) [7].

2.1 The ADoRe algorithm

The first algorithm we illustrate is ADoRe (active disk on request), which combines and extends the RxW algorithm with a C-LOOK scheduling algorithm for disk [8]. The simple algorithm RxW does not meet the requests for data exchange in a ubiquitous environment: this algorithm must be executed as many times as there are requests in the queue in the input to the server.

If we assume that the tail is composed of n elements, the RxW algorithm will initially consider the n RxW products corresponding to the requests and then selecting the one with the highest value. At this point, data corresponding to the selected request are searched in the memory and sent to users who requested them. As a consequence, $n-1$ applications have been queued: these will be subject to a new running of the algorithm, so that the queue is reduced to $n-2$ elements and so on until the data requests are completed. It is unthinkable to adopt a technique of this kind, as it would require a major effort for the server's CPU. Consider then the fact that the queue is not static, but dynamic, i.e. it will be varying over time: if the requirements were initially reduced to $n-1$, others could superpose to them, bringing the queue to an even higher size than the initial one. Regardless of the computing power of the CPU, this algorithm would result in a waste of time, which cannot be affordable by any ubiquitous system.

The ADoRe algorithm overcomes the drawbacks of the RxW. Let us suppose that the queue is usually composed of n elements and K is a natural number constant. The ADoRe sorts n requests in descending order according to the RxW value and is responsible to select the first K requests that have this highest value. The queue is thus reduced to $n-K$ elements. A subsequent implementation of the algorithm will reduce the queue to $n-K$ elements, if other requests have not overlapped in the meantime. ADoRe essentially requires a less number of runs of the scheduling algorithm than those required by the RxW. This means a lower load for the CPU and faster response times to requests (in fact, in this case, data related to sets of K requests are searched on a disk, as a consequence the search optimization on disk is significantly improved, especially when K is high). So if $K = 1$, the ADoRe algorithm is the same as the RxW algorithm. We observe that a peculiarity of ADoRe is represented by the utilization coefficient of the disk (among other things, if the number of elements in the queue is less than K, the algorithm sorts and sends requests to the disk by holding it still busy).

2.2 The Flush algorithm

The good performances of the ADoRe algorithm are due to high disk utilization and optimization of the scheduling of the server waiting queue.

The key idea for the development of FLUSH is to maximize the utilization of the disk. Although FLUSH always uses the C-LOOK as disk scheduling algorithm, it manages differently the user requests to the server node. Each time the disk ends the service for a single request, all the others currently on the queue of the server node flow towards the disk system and are included in its queue, thus becoming subject to the processing of the C-LOOK algorithm. Then, the FLUSH essentially does not employ scheduling algorithms for the input queue of the server. The final result is that long queues are formed in the input of the disk (the so-called scan lists): the C-LOOK algorithm can optimize the data search on disk; as a matter of fact, it now has an overview of all data, which it must search on disk, and not just a partial view of them.

It analyzes the disk from the inside and had learned what are all the data to be taken: it is like as if at one time it had the chance to collect all data required, without returning close to the disk axis, with further radial movements (the reduction of the radial movement is the predominant factor in the optimization of data search speed on the disk) [7].

2.3 The OWeiST algorithm

This algorithm (optimal weighted service time) is different from the previous ones since it can be considered into a category of algorithms that can be denominated 'mixed' [8]. They combine information that is available to the server and the disk, producing a single criterion of scheduling. The obvious motivation to study these algorithms is to determine if the single criterion of scheduling can lead to higher performance than those of the mechanisms that simply combine separate algorithms for scheduling of the disk and the server node. This algorithm tries to improve the performance of the selection process, using the information derived from the queue server and the disk. As we shall see, the research data on disk is different from the C-LOOK and becomes more beneficial as more extensive groups of requests are taken into consideration. Let us suppose to have a number of requests queued to the server, and K is a natural number constant and characterizing the ubiquitous node.

Besides, let Sstart be the last data taken from the disk. The algorithm takes into account all the possible K-uple $(r_1, r_2, \ldots, r_K)$ of requests in the waiting list. Each r_i element of the K-uple refers to the request of a particular data object S_i on the disk, required by R_i clients. Besides, for the K-uple $(r_1, r_2, \ldots, r_K)$, data on disk are supposed to be taken in the respective order $S_1, S_2, \ldots, S_K$, given that the starting position of the head is Sstart. For each K-uple, varying the elements that constitute it and their service order, the time with which information $S_1, S_2, \ldots, S_K$ are taken from the disk varies.

The OWeiST algorithm must, therefore, calculate almost all the possible K-uple and it is the one which shows the least sum of products $R_i x T_{i,i-1}$, where $T_{i,i-1}$ is the time equal to the shift of the head from the position S_{i-1} to the position S_i on disk. The access time $T_{i,i-1}$ takes into consideration the positioning time from the cylinder of the $i-1$th set to the cylinder of the ith set, and the rotational latency necessary for positioning the head on the object i, once the cylinder is reached. The algorithm calculates for every possible permutation of K the above-required sum. On identifying the best K-uple, the requests contained in it will be sent to the queue entry in the disk to be met. After the service for the K-uple is finished, the algorithm is again applied to select a new group of K requests. This process continues until the elements present in the waiting queue are exhausted. Note that if $m < K$ requests queued to the server were present, the algorithm will order them with the same procedure, forming however an m-uple instead of a K-uple. Obviously, by limiting the value of the parameter K, we can control the overhead introduced in the search for the best K-uple [7].

2.4 The RxW/S algorithm

This RxW/S algorithm also belongs to the mixed category. This algorithm is much more simple than the previous one. Given a certain queue of incoming requests to the server, for each element of it, it is estimated the value of the waiting time *W*, the value of the number of users (nodes) linked to the '*i*' request, and the time Sstart, i.e. the time required to position the head on the requested object, starting from the location of the last set of data related to the last fulfilled request. The value of RxW/Sstart,*i* is then calculated for each request in the queue. Among these, the first to be satisfied is the one for which the above-mentioned product is maximum. This algorithm retains some of the features of the algorithm RxW, while taking into account the physical access time to resources on the secondary memory. This algorithm is a one-step algorithm: it is executed every time a request is fulfilled (i.e. requested information are taken from the disk and sent). In this algorithm there is no grouping of requests, because each request is served individually [7].

2.5 Cache memory in a server node

Cache is nothing but a memory of much smaller size compared with a mass storage device, but at the same time it is much faster: its access time is significantly lower than that of a disk. The basic idea is to use the cache to store more recently and more frequently used data to speed up the request of users. Let us see then how to change the characteristics of the algorithms examined earlier, when the ubiquity of the server node includes cache. In particular, algorithms for the management of the cache must be taken into account. Scheduling policies under consideration involve the combined use of algorithms for scheduling such as least recently used (LRU) and least frequently used (LFU), which are considerably efficient and simple at the same time. The combined use of two techniques is more efficient than the use of the same taken individually. It is also more advantageous than other cache management algorithms (such as LRU-K).

2.6 LF-LRU algorithm

This algorithm uses a cache buffer with a given capacity in terms of storing sets of data. The algorithm works by using two order lists: a LRU and a LFU list. The data entering the buffer are placed from time to time at the top of the LRU list. For each set of data that enters the buffer, the algorithm counts the number of references (here reference means the request of that data from one or more users) during the time it is in the buffer. Give a request for a certain set of data, the algorithm first checks if the data requested is already present in the cache. If so, the reference counting for that set is increased and the data is moved from its present position in the LRU list, at the top of it. If the data is not in the cache, it will be searched in the disk, located at the top of the LRU list, with the reference count initialized to 1. The LFU list is an ordered list of all the data in the cache that are arranged in no decreasing order for counting references values.

The LF-LRU policy provides replacement when one data is not found in the cache, it is searched in the disk and replaces in the cache itself, the set of data that has the lowest value of counting references. Among the data indicating the minimum reference counting, the one that leaves the cache is that one which lies at the bottom of the LRU list, namely, that one which for the longest time, among them, was not required. In essence, the cache is able to store inside the most recently and frequently used data. Each time a set of data leaves the data cache, the corresponding reference count is obviously cleared. This mechanism allows the cache to adapt itself to changes in user demand on the nature of data required: it is therefore a kind of adaptive algorithms, which provide that all more frequently required data, among the M available, may change over time. Note that the queue scheduling policies of the requests from the server node remains the same as specified earlier. The only thing that changes is the 'place' where the data desired by users will be searched (in this case first on the cache and then on the disk, while in the previous it is only on the disk).

2.7 LRU-K algorithm

This is an evolution of the LRU algorithm. The LRU algorithm, as mentioned earlier, deletes the data that has not been used for a long time from the list of more recent references. The LRU-K algorithm numbers the requests submitted to the server associated with a natural number n representative of the instant of time when the request was made (the first request received by the server is associated with the number 1, the second to the number 2, the third to 3 and so on). At this point, let us suppose that at a certain instant of time t, the $r_1, r_2, \ldots, r_h$ requests have been made, and let us suppose that q is the generic set of data among the M sets present in the memory.

We define the K-distance $D_t(q,K)$, relative to the set q at time t, as the number of steps to execute 'backward', sweeping the number n starting from h until the K more recent requests of the set q are reached. For example, if 100 requests $r_1, r_2, \ldots, r_{100}$ have arrived at instant t, supposing that requests $r_7, r_9, r_{17}, r_{22}, r_{36}, r_{74}, r_{78}$, and r_{91} are relative to the set q and that K=5, then (being h=100) the K-distance $D_t(q,5)$ will be 100-22, then to reach the first of the last 5 requests of the set q we have to execute 100-22 backward steps sweeping the index *n*. If five requests of the set *q* have not yet been made, its *K*-distance is thought to be infinite.

The LRU-K algorithm has, at instant *t*, a list of request data set in an increasing order according to the *K*-distance value. In this manner it is possible to have a deeper knowledge of which are data that are effectively used by the server (i.e. they are requested), considering the elapsed time since the last *K*-requests of the generic set *q* have been requested [5].

When a request data has not been found, it is searched on the disk, and then it is made available to all users and is loaded into the cache. Because of the limited size of the cache, it is probable that a substitution of data is required. The data that just leave the cache is the one that has the highest *K*-distance. It is obvious that if the parameter *K* grows, a more detailed description of the history of the references at instant *t* is available. It may appear that high values of *K* are associated to an improvement in the performance of the server's response to the ubiquitous user nodes. In reality, there is to consider that the increase of K implies the growth of the running costs of the algorithm, making no sense an implementation of the algorithm to exchange data in ubiquitous environment. The designers generally suggest a value of $K = 2$ as the best compromise between quality and price in the solution of most problems of information exchange [5].

2.8 Considerations on the use of a finite speed transmission channel

The speed of the transmission channel connecting the user node to the server node plays a key role on the performance of information exchange. The basic element to be considered, if the channel has finite speed of propagation is that the data sent to the transmitter of the server node are not sent immediately to clients who requested them, because of the 'slowness' of the transmission channel. In this manner, queues waiting to be sent to users are formed at the input of the transmitter. These queues slow down the server's response and affect its global behaviour.

Three different values of transmission rate for channel were considered: 2, 155 and 620Mbps. From the analysis of the results obtained with low values of transmission rate of the channel (e.g. 2Mbps), it has been observed that the speed of transmission of the channel element is the bottleneck of the server node. The behaviour of FLUSH, supported by LF-LRUs, with different sizes of cache, has been observed and it has been noted that the tendency to decrease the average response time as the cache grows is not valid because of significant delays introduced by channel of transmission: in practice, taking data from the disk or from the cache involves almost no difference, since the delays introduced by the channel transmission override those possibly introduced by an excessive use of the disk.

Conversely, considering the means of transmission at high speed (e.g. 620Mbps or even 155Mbps examined earlier), the behaviour of the average response time of the system is exactly the same as previously considered: the fact of having a high speed transmission channel highlights the importance of having a cache memory. In conclusion, for high-speed channels, the 'bottle-neck' of the server node is given by the disk [7].

3 Context-awareness

We have focused our attention on issues that may make possible the discovery of a service, citing the problem of detection of the users or the knowledge of the environment by ubiquitous devices; all of this constitutes the context, while the capability of the devices to interact in a certain manner in a given situation, that is, of having a knowledge of the environment in which they work, is named 'context-awareness'.

3.1 What is context-awareness?

The term *context* is used to denote a set of information characterizing the state, the situation in which there is an entity, meaning as an entity a person, an object or a place, which is taken into consideration [9–11]. If, for example, a project for a tourist guide is considered, the context will be all information regarding its position on what tourists want to visit, their personal preferences, etc., so the context is not only a set of purely physical information, but personal information, preferences, their knowledge and previous interactions with the system, which can also be viewed as part of the context.

3.2 Possible applications

The theory on the understanding of the context opens the way for economic development and the design of many projects [12,13] based on using hand-held devices, their prototype should ideally be provided with a screen allowing the user interface through special pens or just using fingers, or the ability to access the physical memory devices, such as a CD, as well as to remote resources via the Web, easy ways to communicate with other devices, interface and video, or audio input and output, for instance, based on speech recognition or cameras that enable it to interpret the gestures of a user or to recognize objects and symbols of the surrounding environment. The applications of such devices [11,14,15] are manifold and behind them there is the need that the system has knowledge of the context of the user.

For example, in a museum such a system can provide the visitor (with PDA or similar device) information related to the particular point of the building where it is located, the system recognizes the user's position using different solutions: a bar code, wireless and IR sensors. If in a museum we are in front of an artwork, we can obtain additional information about it and its author. But the tourist guide is not the only possible application. In a hospital, like a museum, the physician as the visitor may have specific information useful associated to the position-location in which they are. An example of application is the 'bus catcher' [16,17], which aims to provide detailed information on public services, such as the location of the means, routes and times of arrival.

Acknowledgements

This chapter was written with the contribution of the following students who attended the lessons of 'Grids and Pervasive Systems' at the faculty of Engineering in the University of Palermo, Italy: Barrale, Cannizzo, Muratore, Scoma, Trapani and Vaste.

Authors would also like to thank Giovanni Pilato for his help in the Italian–English translation.

References

[1] Chakraborty, D., Perich, F., Avancha, S. & Joshi, A., DReggie: semantic service discovery for M-commerce applications. *Workshop on Reliable and Secure Applications in Mobile Environment, Symposium on Reliable Distributed Systems*, New Orleans, LA, October 2001.

[2] Kagal, L., Korolev, V., Avancha, S., Joshi, A., Finin, T. & Yesha, Y., *Highly adaptable infrastructure for service discovery and management in ubiquitous computing*. Technical report, TR CS-01-06, Department of Computer Science and Electrical Engineering, University of Maryland Baltimore County, Baltimore, MD, 2001.

[3] Kagal, L., Korolev, V., Chen, H., Joshi, A. & Finin, T., Project Centaurus: a framework for indoor services mobile services. *Proc. of Int. Workshop on Smart Appliances and Wearable Computing IWSAWC, 21st Int. Conf. on Distributed Computing Systems (ICDCS-21)*, Department of Computer Science and Electrical Engineering, University of Maryland Baltimore County, Baltimore, MD, April 2001.

[4] Dykeman, H., Ammar, M. H. & Wong, J., Scheduling algorithms for videotext systems under broadcast delivery. *Proc. of Int. Conf. of Communications*, pp. 1847–1851, 1996.

[5] Acharya, S. & Muthukrishnan, S., Scheduling on-demand broadcasts: new metrics and algorithms. *Proc. of the 4th Ann. ACM/IEEE Int. Conf.*

on Mobile Computing and Networking, October 25–30, Dallas, TX, pp. 43–54, 1998.

[6] Aksoy, D. & Franklin, M. RxW: a scheduling approach for large-scale on-demand data broadcast. *IEEE/ACM Transactions on Networking*, **7(6)**, pp. 846–860, 1999.

[7] Triantafillou, P., Harpantidou, R. & Paterakis, M., High performance data broadcasting systems. *Mobile Networks and Applications*, **7(4)**, pp. 279–290, 2002.

[8] Triantafillou, P., Harpantidou, R. & Paterakis, M. High performance data broadcasting: a comprehensive systems' perspective. *Lecture Notes in Computer Science*, pp. 79–90, 1987/2001.

[9] O'Hare, P.T., O'Hare, G.M.P. & Lowen, T.D., Far and a way: context sensitive service delivery through mobile lightweight PDA hosted agents. *Proc. of the 15th Int. Florida Artificial Intelligence Research Society Conf.*, May 14–16, pp.13–17, 2002.

[10] Abowd, G.D., Dey, A.K., Brown, P.J., Davies, N., Smith, M. & Steggles, P., Towards a better understanding of context and context-awareness. *Proc. of the 1st Int. Symp. on Handheld and Ubiquitous Computing*, September 27–29, Karlsruhe, Germany, pp. 304–307, 1999.

[11] Anind K. Dey, A.K., Salber, D., Abowd, G.D. & Futakawa, M., The conference assistant: combining context-awareness with wearable computing. *Wearable Computers, IEEE Int. Symp., 3rd Int. Symp. on Wearable Computers (ISWC'99)*, October 18–19, IEEE Computer Society: Washington, DC, p. 21, 1999.

[12] Huang, A.C., Ling, B.C. & Ponnekanti, S., Pervasive computing: what is it good for?. *Proc. of the 1st ACM Int. Workshop on Data Engineering for Wireless and Mobile Access*, August 20–20, Seattle, Washington, DC, pp. 84–91, 1999.

[13] Shafer, S., Ubiquitous computing and the EasyLiving Project. *Invited Presentation at 40th Anniversary Symposium of Osaka Electro-Communications University*, November 2001.

[14] Schilit, B., Adams, N. & Want, R., Context-aware computing applications. *Proc. of Workshop on Mobile Computing Systems and Applications*, December 8–9, pp. 85–90, 1994.

[15] Ward, A., Jones, A. & Hopper, A., A new location technique for the active office. *Personal Communications IEEE*, **4(5)**, pp. 42–47, 1997.

[16] Strahan, R., Muldoon, C., O'Hare, G.M.P., Bertolotto, M. & Collier, R.W., An agent-based architecture for wireless bus travel assistants. *Proc. 2nd Int. Workshop on Wireless Information Systems, WIS 2003, in conjunction with ICEIS 2003,* Angers, France, April, pp. 54–62, 2003.

[17] Bertolotto, M., O'Hare, G., Strahan, R., Brophy, A., Martin, A. & McLoughlin, E., Bus catcher: a context sensitive prototype system for public transportation users. *Proc. of the 3rd Int. Conf. on Web Information Systems Engineering (Workshops)*, pp. 64–72, 2002.

Index